Forensic Science: A Very Short Introduction

VERY SHORT INTRODUCTIONS are for anyone wanting a stimulating and accessible way in to a new subject. They are written by experts, and have been published in more than 25 languages worldwide.

The series began in 1995, and now represents a wide variety of topics in history, philosophy, religion, science, and the humanities. The VSI library now contains over 200 volumes—a Very Short Introduction to everything from ancient Egypt and Indian philosophy to conceptual art and cosmology—and will continue to grow to a library of around 300 titles.

Very Short Introductions available now:

AFRICAN HISTORY John Parker and Richard Rathbone
AMERICAN POLITICAL PARTIES AND ELECTIONS L. Sandy Maisel
THE AMERICAN PRESIDENCY Charles O. Jones
ANARCHISM Colin Ward
ANCIENT EGYPT Ian Shaw
ANCIENT PHILOSOPHY Julia Annas
ANCIENT WARFARE Harry Sidebottom
ANGLICANISM Mark Chapman
THE ANGLO-SAXON AGE John Blair
ANIMAL RIGHTS David DeGrazia
ANTISEMITISM Steven Beller
THE APOCRYPHAL GOSPELS Paul Foster
ARCHAEOLOGY Paul Bahn
ARCHITECTURE Andrew Ballantyne
ARISTOTLE Jonathan Barnes
ART HISTORY Dana Arnold
ART THEORY Cynthia Freeland
ATHEISM Julian Baggini
AUGUSTINE Henry Chadwick
AUTISM Uta Frith
BARTHES Jonathan Culler
BESTSELLERS John Sutherland
THE BIBLE John Riches
BIBLICAL ARCHEOLOGY Eric H. Cline
BIOGRAPHY Hermione Lee
THE BOOK OF MORMON Terryl Givens
THE BRAIN Michael O'Shea
BRITISH POLITICS Anthony Wright
BUDDHA Michael Carrithers
BUDDHISM Damien Keown
BUDDHIST ETHICS Damien Keown
CAPITALISM James Fulcher
CATHOLICISM Gerald O'Collins
THE CELTS Barry Cunliffe
CHAOS Leonard Smith
CHOICE THEORY Michael Allingham
CHRISTIAN ART Beth Williamson
CHRISTIANITY Linda Woodhead
CITIZENSHIP Richard Bellamy
CLASSICAL MYTHOLOGY Helen Morales
CLASSICS Mary Beard and John Henderson
CLAUSEWITZ Michael Howard
THE COLD WAR Robert McMahon
COMMUNISM Leslie Holmes
CONSCIOUSNESS Susan Blackmore
CONTEMPORARY ART Julian Stallabrass
CONTINENTAL PHILOSOPHY Simon Critchley
COSMOLOGY Peter Coles
THE CRUSADES Christopher Tyerman
CRYPTOGRAPHY Fred Piper and Sean Murphy
DADA AND SURREALISM David Hopkins
DARWIN Jonathan Howard
THE DEAD SEA SCROLLS Timothy Lim
DEMOCRACY Bernard Crick
DESERTS Nick Middleton
DESCARTES Tom Sorell
DESIGN John Heskett
DINOSAURS David Norman
DOCUMENTARY FILM Patricia Aufderheide
DREAMING J. Allan Hobson
DRUGS Leslie Iversen

THE EARTH Martin Redfern
ECONOMICS Partha Dasgupta
EGYPTIAN MYTH Geraldine Pinch
EIGHTEENTH-CENTURY BRITAIN
Paul Langford
THE ELEMENTS Philip Ball
EMOTION Dylan Evans
EMPIRE Stephen Howe
ENGELS Terrell Carver
EPIDEMIOLOGY Roldolfo Saracci
ETHICS Simon Blackburn
THE EUROPEAN UNION
John Pinder and Simon Usherwood
EVOLUTION
Brian and Deborah Charlesworth
EXISTENTIALISM Thomas Flynn
FASCISM Kevin Passmore
FEMINISM Margaret Walters
FASHION Rebecca Arnold
THE FIRST WORLD WAR
Michael Howard
FORENSIC SCIENCE Jim Fraser
FOSSILS Keith Thomson
FOUCAULT Gary Gutting
FREE WILL Thomas Pink
FREE SPEECH Nigel Warburton
THE FRENCH REVOLUTION
William Doyle
FREUD Anthony Storr
FUNDAMENTALISM
Malise Ruthven
GALAXIES John Gribbin
GALILEO Stillman Drake
GAME THEORY Ken Binmore
GANDHI Bhikhu Parekh
GEOGRAPHY
John Matthews and David Herbert
GEOPOLITICS Klaus Dodds
GERMAN LITERATURE Nicholas Boyle
GLOBAL CATASTROPHES Bill McGuire
GLOBAL WARMING Mark Maslin
GLOBALIZATION Manfred Steger
THE GREAT DEPRESSION AND THE
NEW DEAL Eric Rauchway
HABERMAS James Gordon Finlayson
HEGEL Peter Singer
HEIDEGGER Michael Inwood
HIEROGLYPHS Penelope Wilson
HINDUISM Kim Knott
HISTORY John H. Arnold
THE HISTORY OF ASTRONOMY
Michael Hoskin
THE HISTORY OF LIFE
Michael Benton
THE HISTORY OF MEDICINE
William Bynum
THE HISTORY OF TIME
Leofranc Holford-Strevens
HIV/AIDS Alan Whiteside
HOBBES Richard Tuck
HUMAN EVOLUTION Bernard Wood
HUMAN RIGHTS Andrew Clapham
HUME A. J. Ayer
IDEOLOGY Michael Freeden
INDIAN PHILOSOPHY Sue Hamilton
INTELLIGENCE Ian J. Deary
INTERNATIONAL MIGRATION
Khalid Koser
INTERNATIONAL RELATIONS
Paul Wilkinson
ISLAM Malise Ruthven
ISLAMIC HISTORY Adam Silverstein
JOURNALISM Ian Hargreaves
JUDAISM Norman Solomon
JUNG Anthony Stevens
KABBALAH Joseph Dan
KAFKA Ritchie Robertson
KANT Roger Scruton
KIERKEGAARD Patrick Gardiner
THE KORAN Michael Cook
LAW Raymond Wacks
LINCOLN Allen C. Guelzo
LINGUISTICS Peter Matthews
LITERARY THEORY Jonathan Culler
LOCKE John Dunn
LOGIC Graham Priest
MACHIAVELLI Quentin Skinner
THE MARQUIS DE SADE
John Phillips
MARX Peter Singer
MATHEMATICS Timothy Gowers
THE MEANING OF LIFE
Terry Eagleton
MEDICAL ETHICS Tony Hope
MEDIEVAL BRITAIN
John Gillingham and Ralph A. Griffiths
MEMORY Jonathan K. Foster
MODERN ART David Cottington
MODERN CHINA Rana Mitter
MODERN IRELAND Senia Pašeta

MODERN JAPAN
Christopher Goto-Jones
MOLECULES Philip Ball
MORMONISM
Richard Lyman Bushman
MUSIC Nicholas Cook
MYTH Robert A. Segal
NATIONALISM Steven Grosby
NELSON MANDELA Elleke Boehmer
NEOLIBERALISM
Manfred Steger and Ravi Roy
THE NEW TESTAMENT AS LITERATURE Kyle Keefer
NEWTON Robert Iliffe
NIETZSCHE Michael Tanner
NINETEENTH-CENTURY BRITAIN
Christopher Harvie and H. C. G. Matthew
THE NORMAN CONQUEST
George Garnett
NORTHERN IRELAND
Marc Mulholland
NOTHING Frank Close
NUCLEAR WEAPONS
Joseph M. Siracusa
THE OLD TESTAMENT
Michael D. Coogan
PARTICLE PHYSICS Frank Close
PAUL E. P. Sanders
PHILOSOPHY Edward Craig
PHILOSOPHY OF LAW
Raymond Wacks
PHILOSOPHY OF SCIENCE
Samir Okasha
PHOTOGRAPHY Steve Edwards
PLATO Julia Annas
POLITICAL PHILOSOPHY David Miller
POLITICS Kenneth Minogue
POSTCOLONIALISM Robert Young
POSTMODERNISM Christopher Butler
POSTSTRUCTURALISM
Catherine Belsey
PREHISTORY Chris Gosden
PRESOCRATIC PHILOSOPHY
Catherine Osborne
PRIVACY Raymond Wacks
PURITANISM Francis J. Bremer
PSYCHIATRY Tom Burns
PSYCHOLOGY
Gillian Butler and Freda McManus
THE QUAKERS Pink Dandelion
QUANTUM THEORY
John Polkinghorne
RACISM Ali Rattansi
THE REFORMATION Peter Marshall
RELATIVITY Russell Stannard
RELIGION IN AMERICA Timothy Beal
THE REAGAN REVOLUTION Gil Troy
THE RENAISSANCE Jerry Brotton
RENAISSANCE ART
Geraldine A. Johnson
ROMAN BRITAIN Peter Salway
THE ROMAN EMPIRE Christopher Kelly
ROUSSEAU Robert Wokler
RUSSELL A. C. Grayling
RUSSIAN LITERATURE
Catriona Kelly
THE RUSSIAN REVOLUTION
S. A. Smith
SCHIZOPHRENIA
Chris Frith and Eve Johnstone
SCHOPENHAUER
Christopher Janaway
SCIENCE AND RELIGION
Thomas Dixon
SCOTLAND Rab Houston
SEXUALITY Véronique Mottier
SHAKESPEARE Germaine Greer
SIKHISM Eleanor Nesbitt
SOCIAL AND CULTURAL ANTHROPOLOGY
John Monaghan and Peter Just
SOCIALISM Michael Newman
SOCIOLOGY Steve Bruce
SOCRATES C. C. W. Taylor
THE SOVIET UNION Stephen Lovell
THE SPANISH CIVIL WAR
Helen Graham
SPINOZA Roger Scruton
STATISTICS David J. Hand
STUART BRITAIN John Morrill
SUPERCONDUCTIVITY
Stephen Blundell
TERRORISM Charles Townshend
THEOLOGY David F. Ford
THOMAS AQUINAS Fergus Kerr
TRAGEDY Adrian Poole
THE TUDORS John Guy
TWENTIETH-CENTURY BRITAIN
Kenneth O. Morgan

THE UNITED NATIONS
Jussi M. Hanhimäki
THE VIKINGS Julian Richards
WITTGENSTEIN A. C. Grayling
WORLD MUSIC Philip Bohlman
THE WORLD TRADE ORGANIZATION
Amrita Narlikar
WRITING AND SCRIPT
Andrew Robinson

Available soon:

PROGRESSIVISM Walter Nugent
INFORMATION Luciano Floridi
THE LAWS OF THERMODYNAMICS
Peter Atkins
INNOVATION
Mark Dodgson and David Gann
WITCHCRAFT
Malcolm Gaskill

For more information visit our web site
www.oup.co.uk/general/vsi/

Jim Fraser

FORENSIC SCIENCE

A Very Short Introduction

OXFORD
UNIVERSITY PRESS

OXFORD
UNIVERSITY PRESS

Great Clarendon Street, Oxford OX2 6DP

Oxford University Press is a department of the University of Oxford.
It furthers the University's objective of excellence in research, scholarship,
and education by publishing worldwide in

Oxford New York

Auckland Cape Town Dar es Salaam Hong Kong Karachi
Kuala Lumpur Madrid Melbourne Mexico City Nairobi
New Delhi Shanghai Taipei Toronto

With offices in

Argentina Austria Brazil Chile Czech Republic France Greece
Guatemala Hungary Italy Japan Poland Portugal Singapore
South Korea Switzerland Thailand Turkey Ukraine Vietnam

Published in the United States
by Oxford University Press Inc., New York

First published 2010

British Library Cataloguing in Publication Data

Data available

Library of Congress Cataloging in Publication Data

Data available

Typeset by SPI Publisher Services, Pondicherry, India
Printed in Great Britain by
Ashford Colour Press Ltd, Gosport, Hampshire

ISBN 978-0-19-955805-6

1 3 5 7 9 10 8 6 4 2

Contents

Preface and acknowledgements

There is more interest in forensic science now than at any previous time in its history. There are more students studying 'forensic' courses in the UK than ever before and there is a seemingly endless list of TV dramas that are testimony to huge popular interest in the subject. In real life, forensic science attracts enormous media attention in high-profile cases such as the deaths of Damilola Taylor, Jill Dando, and Rachel Nickel. More importantly, forensic science provides 'leads' in police investigations and evidence for prosecutions that were previously unimaginable. Despite this, understanding of forensic science is poor even amongst those, such as lawyers and police officers, who are required to use it as well as others such as politicians and journalists. Public understanding of the subject is largely based on TV shows, such as *CSI* (*Crime Scene Investigation*), which use hi-tech imagery for dramatic effect at the expense of understanding of an increasingly important part of the criminal justice process. There is even the so-called 'CSI effect' – that expectations and misconceptions about forensic science on the part of the public may have adverse influence on jury decisions.

Dramatic scientific breakthroughs, particularly the discovery of DNA profiling, in the past 20 years or so have revolutionized forensic science. Evidence can be obtained from microscopic traces of body fluids, drugs, and explosives of sufficient quality for it to be pivotal in an investigation or trial. There has been a parallel revolution in how the police investigate crime. It is probably more effective, faster, and more reliable to investigate the crimes that affect us most (burglary, car theft, and suchlike) using DNA and fingerprints than by any other means. In major crime, such as homicide, forensic scientists have moved from being backroom boffins to the forefront of international investigations. Forensic science is now firmly embedded in the criminal justice agenda since it can answer investigative questions in many instances better than any other means available. It is a complex activity at the interface of science and law. Forensic science is not a discipline in its own right, but engages many disciplines such as chemistry, molecular biology, and engineering, though it has a number of distinctive features. Whilst rooted in science, it is an intensely practical activity that deals with real-world issues: explosions, blood spatters, bodies, and stolen cars. Complex scientific findings must be weighed carefully and dispassionately, and communicated with clarity, simplicity, and precision to police, lawyers, jurors, and the judiciary. Forensic science encounters all aspects of human behaviour. The famous headline 'all human life is here' fits forensic science very well: the plain stupid (the killers who panicked and re-buried a body for the third time in a flower bed in a graveyard); the unlucky (the man who wrote an anonymous threatening letter to the chairman of a London football club on paper with invisible indented impressions of his name and home address); to the cold and frighteningly malevolent – serial sexual offenders and killers who plan and fantasize about their crime throughout the course of their life (Anthoni Imiela and Robert Black). In short, forensic science matters because the link to everyday life (and death) is more direct, tangible, and visible. But forensic science does not have all the answers. In some instances, it has no answers at all (for example in the Michael Stone case), and in some cases it fails spectacularly and worryingly for reasons that are not always clear, for example in the Jill Dando case. Forensic science is also regarded ambivalently by some (as is science by the public in general) and by others as a source of injustice. The arguments of the latter are rarely well informed in my experience, but I will explore some of these issues in this book.

It would be impossible to do justice to all areas of forensic science in a book of this type and length, so I have necessarily had to select some things and exclude others. Whole areas of forensic science are completely absent: toxicology, crash investigation, computer forensics, document examination, and others are dealt with superficially or in passing. In making this selection, I have attempted to identify the central issues of forensic science, such as identification and evidence evaluation, and its main procedures and mechanisms, such as continuity of evidence (chain of custody in the USA and many other countries) and minimising contamination. Many of the cases I have used as illustrations come from direct personal involvement and memory. I have not provided detailed information in every case as this is rarely necessary to gain an understanding, but in some instances the full details are already well publicized. It is my contention that you do not need to know the details of every area of forensic science to know the nature of forensic science. I will leave the reader to judge the success or otherwise of my efforts.

Although science uses more or less universal terminology, that used in policing and the law varies considerably even to the extent that the same word can mean different things in different jurisdictions. For example, the document containing forensic science evidence presented to the courts in England is called a 'statement', whereas the equivalent document in Scots Law is called a 'joint report' and a statement means something else. In Scotland, items produced in evidence are called 'productions', whereas in England, the USA, Australia, and many other countries they are called 'exhibits'. This is a constant problem when discussing or teaching forensic science. To overcome this, I have decided to abandon all attempts to be legally precise except where essential and have used common-sense terminology such as item (instead of production or exhibit) or report (instead of 'joint report' or 'statement'). None of these infringements should impede understanding of the subject. The chapters generally follow the chronological flow of how forensic science interacts with the criminal law – incident, investigation, and laboratory analysis – from crime scene to court.

Finally, a word on those 'CSI' or 'eureka' moments – when the scientist 'cracks' the case with a piece of brilliant incisiveness and basks in the admiration of her colleagues. Yes, they happen, but far less frequently than TV dramas would have you believe. Perhaps five or six times in a long career this might occur. In truth, most cases are solved by a combination of systematic investigation by a range of professionals (police officers, scientists, pathologists, CSIs), good teamwork, effective leadership, hard work, and some luck. I hope this comes across from the text.

I am indebted to many for their support in the writing of this book: the initial reviewers, colleagues, friends, and all who provided advice, critical comment, and images. I wish to thank them all (in alphabetical order): Sarah Cresswell, Peter Gill, Jim Govan, Isobel Hamilton, Max Houck, Anya Hunt, Lester Knibb, Adrian Linacre, Terry Napier, Niamh NicDaeid, James Robertson, Derek Scrimger, Nigel Watson, Robin Williams. I would also like to thank Latha Menon for her enthusiasm in commissioning the project and Emma Marchant for seeing it through with me. Finally, special thanks to my partner Celia and son Robbie for their enduring patience when I should have been paying more attention to them and not locked in my study.

List of illustrations

List of tables

Chapter 1
What is forensic science?

> The bloodstains looked like the scattered fragments of a mysterious pattern – a last message, a warning, the writing on the wall.
>
> Alec Ross, *The Rest is Noise: Listening to the Twentieth Century*

These were the words of Klaus Mann (the son of Thomas Mann) following his discovery of the corpse of his friend and former lover Ricki Hallgarten who had shot himself through the heart. Paul Kirk expressed a similar sentiment in even more detail and in more utilitarian terms:

> Wherever he steps, whatever he touches, whatever he leaves, even unconsciously, will serve as a silent witness against him. Not only his fingerprints or his footprints, but his hair, the fibres from his clothes, the glass he breaks, the tool mark he leaves, the paint he scratches, the blood or semen he deposits or collects. All of these and more, bear mute witness against him. This is evidence that does not forget. It is not confused by the excitement of the moment. It is not absent because human witnesses are. It is factual evidence. Physical evidence cannot be wrong, it cannot perjure itself, it cannot be wholly absent. Only human failure to find it, study and understand it, can diminish its value.

Kirk replaces Mann's lyrical symbolism with anthropocentrism. Not only is there a story to be told but, according to Kirk, one cannot fail to read it. This is what is referred to by my colleague Robin Williams as the 'forensic imaginary' – the conviction that all such events are knowable and can be reconstructed from forensic evidence, that there is always a decipherable last message from the victim and evidence from the perpetrator; the 'signature' of the killer. Mann considers the bloodstains to be not just symbols of violence but a 'text' that can be read and interpreted, and Kirk makes it clear that we cannot fail to do so.

The most influential thinker in forensic science was Edmond Locard (1877–1976), who almost certainly prompted Kirk's comments above. Locard established the first police scientific laboratory for investigating crime scenes in Lyon, France, in 1910. He also set out what many consider to be the fundamental basis and guiding principle of forensic science. This is most frequently formulated as 'every contact leaves a trace', although Locard never used these exact words. Directly or by implication, the message that is taught to police officers and the masses of new forensic science students is that these views represent reality: that there will always be evidence about such events, and ultimately that all things can be known about a crime or a criminal. Only failure on our part as humans can usurp this aim. Also, that this evidence is dispassionate, objective: not only will we know things, there will only be one version of the truth (and therefore no disputes). And we have the last laugh since all this can happen without the criminal even knowing.

From my experience of forensic science, it is difficult to imagine a situation that is much further from reality. Locard's principle as it is usually described is not a scientific theory because it cannot be tested by scientific means, and it cannot predict in the way that scientific laws such as gravity or electromagnetism can. Nor could it be described as a model of the world – we would need much more evidence than we actually have to assert this. It is more a principle based on a thought experiment. Like other 'scientific' principles, for example the cosmological principle, which makes certain simplifying (but untrue) assumptions about the distribution of matter in the universe, the point of it is to help us think about things when we have little or no data to go on. What we do know is that research supports Locard's assertions in part but that there are also limitations to the application of these concepts. The flawed assumption is that once evidence is transferred it remains in place, because we know that this is not the case. Generally speaking, such evidence will be lost and often very quickly, perhaps a few hours after the event, as illustrated in Figure 1. We can therefore put forward as a genuine scientific theory, one that can be tested on the basis of empirical evidence, the concept of transfer and persistence. For example, when items of clothing come into contact, fibres will be transferred from each to the other and then gradually lost.

We have perhaps been a little hard on Locard and Kirk. So let's return to this thought experiment and imagine a world in which

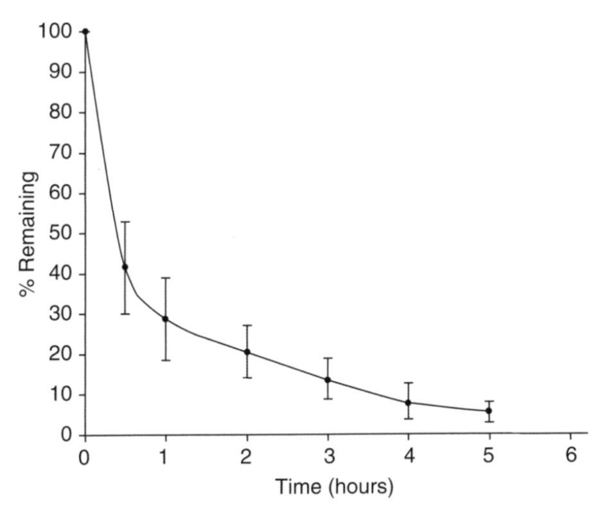

1. Loss of fibres from the surface of the skin. This illustrates a pattern which is typical of fibre loss from many different surfaces. After five hours around 95% of the evidence has been lost

things are constantly being transferred, and as we now know also lost. I sit on a fabric-covered seat on a train reading a book. Fibres from my clothing are transferred to the seat and from the seat to my clothing. When I arrive in my office, some fibres which remain on my clothing from the train seat will be transferred to my office seat. So far so good, this is not too complicated, so let's continue with the experiment. Also being transferred to the train seat were fibres from my home environment, from upholstery, carpets, the clothing of my family, and perhaps hairs from pets. And on the train seat, in addition to fibres from passengers will be fibres from their homes, some of which will transfer to my clothing and perhaps to my office seat. The situation is now rather complicated. There are fibres in my office from people on the train whom I have never been in contact with and have never been in my office (although most of these will be lost on the walk from the station). There may be fibres from things in my home in other people's offices (who were also on the train). All of these fibres will be mass produced so none of them is unique. It should be clear now that finding fibres that match someone's clothing in my office does not mean that that person has been in my office. In fact, it does not even mean that they have come from that person's clothing. To make sense of any fibres that are found, we need to bring in some more up-to-date concepts in forensic work such as primary (direct) and secondary (indirect) transfer. The fibres transferred from my clothing to the train seat (and the reverse) are due to direct transfer. The fibres from other people's clothing on my office seat are from indirect transfer. So whilst every (direct) contact may leave a trace, traces may also be transferred which are not due to (direct) contact. Forming a view as to whether traces are direct or indirect contact needs a great deal more information, which we will explore in subsequent chapters. It should now be obvious that making sense of this requires the inevitable involvement of fallible humans, uncertain information, scientific tests that have inherent error rates, and subjective interpretation of test results. The final twist to this tale is that all of these activities, examinations, and interpretations must comply with the law and legal procedure.

This takes science from the laboratory directly into a very different world in which the interpretation of the scientific evidence may depend on the law. For although science is essentially universal – it is the same in Glasgow, New York, and Beijing – the law is local, sometimes astonishingly so. Furthermore, in common law systems such as those in the UK, USA, Australia, and Canada, the rules of evidence constrain what can be said and done in court, including what scientific or expert evidence can be presented. The law decides for itself what can and cannot be heard. And fundamental to the common law (or adversarial) process is the notion of argument: that there is inherently more than one viewpoint, position, or interpretation to any set of facts. The law is the final nail in the coffin for Kirk and Locard and any grand vision of uniqueness, objectivity, and infallibility. But we should acknowledge their originality, creative imagination, insight, and the influence they have had in inspiring scientists to develop more rigorous empirically based theories.

So what is forensic science? Definitions are not helpful here as at best they usually suggest a connection or interaction between law and science but do not provide any insight into the complexities or limitations of this odd relationship. From my perspective, it is better to describe than define. For me, forensic science is the investigation, explanation, and evaluation of events of legal relevance including the identity, origin, and life history of humans, materials (e.g. paint, plastics), substances (e.g. drugs and poisons), and artefacts (e.g. clothing, shoes). This is done using scientific techniques or methodologies which allow us to describe, infer, and reconstruct events. The basis of the reconstruction is the analysis and evaluation of indirect fragmentary physical evidence (what remains of the traces) and relevant information. From these facts, when established to some pre-determined legal standard, the law infers behaviour, motivation, and criminal intent. In short, forensic science answers the central questions in a criminal investigation: who, what, where, when, how, and why? Answers to these questions include the identity of the criminal or victim using DNA or fingerprints, what type of shoe left the mark at the crime scene, the sequence of events that led to a death as established by bloodstain pattern analysis, where a shot was fired from, or how a fire has started from a scene investigation and why it burned so fiercely from analysis of flammable liquids. We will consider many of these issues in more detail in subsequent chapters, describing the processes involved, the methods of analysis, how the evidence is interpreted, and ultimately how it is presented in court.

Chapter 2
Investigating crime

Forensic science takes place within the complexities and procedures of the law, its timescales, pressures, and contingencies, the drama of a criminal investigation and occasionally in the intense glare of the media. This chapter explains how the police investigate crime, setting out the procedures and principles involved and how forensic science provides answers to the important questions we identified at the end of Chapter 1. Science is the best means we have of describing and understanding the physical world, and forensic science has fundamentally changed the way the police investigate crime, legally, procedurally, and conceptually, because it can answer many of these important questions faster and more objectively than by any other means.

A crime is a breach of the criminal law that requires two elements: a mental element (*mens rea* – a guilty mind) and a physical element (*actus reus* – the guilty act). It is not possible to directly establish *mens rea* using physical science – this must be inferred from the activities and behaviour of the accused. To commit a crime, the accused must have the intent to do wrong and to carry out the act. A criminal investigation is a search for information with the aim of bringing an offender to justice which is achieved by reconstructing the events leading up to, during, and sometimes after the crime. This is done by gathering facts and information, speaking to witnesses, recovering documents, examining CCTV footage, and carrying out scientific examinations, with the aim of answering the most important questions: who?, what?, why?, when?, where?, how? To do this effectively requires skill, resources, and time. Complex investigations such as serial offences and homicides require considerable coordination and planning both during the investigation and trial.

In the UK and most Westernized countries, the police have systems and procedures in place to help achieve their aims. The modern approach to investigating major crime in the UK followed after serious problems in the investigation of the 'Yorkshire ripper' case in the 1980s. In pre-computer days, the police were overwhelmed by paper records and therefore unable to link together important findings from thousands of separate 'leads' or track the outcomes of various aspects of the investigation. This undoubtedly delayed the arrest and prosecution of Peter Sutcliffe, who was subsequently convicted. A review carried out following the trial recommended major changes in procedure that included the use of standard operating procedures in major inquiries, introduction of computer systems (HOLMES – Home Office Large Major Enquiry System), and a new specialist scientific role in complex enquiries to ensure that forensic science is focused on the primary objectives of the investigation. Although computer systems are used to support crime investigation all around the world, for example the Australian Federal Police use PROMIS (Police Online Management Information System), I am not aware of any other country which uses a single common system such as HOLMES. Most serious crimes will be investigated by a team of detectives and specialist police officers led by a Senior Investigating Officer (SIO) – an experienced and trained detective.

Forensic science also became a matter of significance following a number of high-profile enquiries in England in the 1990s. One of these cases, the 'Birmingham Six', who were wrongly convicted of terrorist offences, included specific failings in the scientific aspects of the investigation. Following a Royal Commission which considered this case and others, radical changes in English law altered how police deal with accused persons, go about their investigations, and are accountable for their actions. These and other changes have led the police to take an approach which is based on the systematic elimination of individuals from an investigation (Trace, Interview, Eliminate – TIE) until one person is left who may be the offender. Today, using computers and standardized processes, the police use the principles of Trace, Interview, and Eliminate to follow specific lines of inquiry that lead to solving the crime. Typical lines of inquiry might include:

- Who was the driver of the car that was last seen leaving the scene of incident?
- When was the deceased last seen?
- Are there any witnesses with information in neighbouring properties? (House-to-house inquires or canvassing)
- Does anyone living in the area have a criminal record of relevance to the inquiry, e.g. sex offenders, burglars?

The TIE process requires the investigator to identify the main lines of inquiry by identifying the key questions to be answered in the case. For example: Who is the deceased person? What took place? Where did the offender and victim meet? When was the victim last seen alive? Why did the offender act the way they did? How was the body disposed of? All relevant lines of inquiry are identified and logged on the computer specially developed for this purpose – HOLMES. This is then used to aid prioritizing of the most important inquiries, the allocation of tasks to specific officers, and the recording of the outcomes from these tasks. All information is then stored in a single database which can be interrogated and used to develop and test hypotheses. HOLMES also helps identify facts and witnesses which corroborate each other, as well as inconsistencies, for example in witness accounts, which can then be re-investigated. If we take a homicide as an example, there will be a number of individuals who will be high on the list for TIE, such as:

- individuals who had access to the scene at the time of the crime;
- known associates of the victim or those living locally;
- individuals with previous convictions, especially for violence;
- those with physical characteristics similar to the suspected offender;
- owners of vehicles of the same type as that known or suspected to be involved.

The HOLMES system also records how someone is eliminated to ensure that nothing and no one is overlooked, as happened in the 'Yorkshire ripper' case. It is notable that the 'number one' method of elimination is 'forensic', that is, this means of elimination is considered to be highly desirable due to its objectivity and reliability compared to others. Alibis based on eyewitnesses, friends, associates, and even independent witnesses all have their flaws, but elimination by forensic science is essentially cast-iron. The complete HOLMES elimination criteria are:

1) Forensic, e.g. the DNA or fingerprints of the individual do not match those known or believed to be from the offender.
2) Description: the physical appearance of an individual does not match that of the offender as known or described by witnesses.
3) Independent witness (alibi). An elimination which is due to the evidence of an independent witness and provides the subject with an alibi.
4) Associate or relative (alibi). An elimination which is based on evidence of an associate or relative of the subject. This is inherently less reliable that the evidence of an independent witness.
5) Spouse or common law relationship (alibi). An elimination based on the evidence of a spouse. This is the least reliable form of alibi.
6) Not eliminated. The subject has not been eliminated from the inquiry and therefore remains of interest to the police. This does not necessarily mean the individual is a suspect, although in certain circumstances (such as incriminating evidence coming to light) they may become so.

This methodology is used only for major and serious crime and requires considerable skill and resources. The TIE process, when applied systematically, gradually accumulates vast amounts of information of increasingly fine granularity on the identity, behaviour, activities, relationships, and history of individuals. This includes information such as where they work, who they go drinking with, what their hobbies are, who they are having affairs with, who is in debt, who has suddenly come into money, who has recently had a dispute or fight. Such inquiries will also reveal relationships between people, and places and objects of interest to the inquiry such as vehicles or crime scenes. Much of this information is of no interest to the police as they will be focused on what is required to be established to detect and then prosecute the particular offence.

Here we come to another 'imaginary' which we could perhaps call the 'investigative imaginary'. This presupposes that all crime will yield to the scrutiny of an investigation and that such events can always be reconstructed, resolved, and closed. Although this is the case in many instances – in fact, for the vast majority of major offences such as homicide – it is by no means certain. The police do not find all the information they seek despite repeated efforts to do so and some that is found remains ambiguous. Despite the public perceptions and the expectations of many, not all things are knowable. The analogy of investigation most frequently used is that of a jigsaw puzzle – so long as you have the pieces, only time and persistence are required to solve the puzzle. This is a poor analogy for such a complex event as a major crime for a number of reasons. Even when the crime is solved and prosecuted, the puzzle is rarely complete. Nor does it need to be, since the law does not expect the answer to every question but proof of the key facts and elimination of reasonable alternatives. And these facts will be rendered in an adversarial legal process in which there are always two sides to any story: there is always an alternative explanation, always some missing piece of the problem. Investigations and trials are contingent, restricted by the rules of evidence and procedure and limited by time and resources. A trial is only a search for the truth in this very narrow sense.

Investigation of many different types of crime can be supported or resolved by a wide range of forensic analyses that can provide the answers to who, what, why, when, where, and how. Certain types of forensic evidence tend to be associated with particular types of crime, although this is not always the case. Table 1 gives an indication of the types of forensic evidence that can answer the main investigative questions and summarizes many of the issues we will explore in subsequent chapters. DNA and fingerprints are the two main methods of identifying people, alive or dead, and provide evidence that is considered conclusive by the courts. Marks, including shoe marks and tool marks, can identify the objects concerned and link them to crime scenes (and indirectly to the individuals who possess them), as well as linking crime scenes together as evidence or intelligence.

Some cases require comparatively straightforward questions to be answered: do the tablets seized by the police contain amphetamine? This can be done by standard methods of chemical analysis. Others ask much more complex questions: what was the sequence of events leading to a death? Most cases will be solved by combinations of evidence, for example linking an individual to a mobile phone using DNA or fingerprints and cell site analysis (which can track mobile phone signals) to establish where the individual was when the phone was being used. The detailed context of the case will determine which questions will be asked by the police and therefore which scientific tests will be carried out. Although certain types of evidence are better than others in a general sense, it is the case context that will determine its true value to an investigation. The specific details of the incident – timings, locations, other evidence – and alternative explanations must be blended to form the final evaluation of the significance of the evidence.

Table 1. Investigative uses of forensic science

Investigative use	Technique/Examination	Examples
Identifying people	DNA, fingerprints	Victims, suspects, witnesses, body parts
Identifying objects	Shoe marks, tool marks, fabric marks, cartridge cases	Establishing shoe make and model from sole pattern or tool type from general characteristics of a mark
Identifying materials and substances	Drugs analysis, paint analysis, flammable liquids	Tablets seized in drugs trafficking, accelerants used in arson
Associating people	Blood, hairs, fibres, body fluids, DNA	Connecting victims with suspects in a wide range of offences
Associating objects	Physical fits, marks (shoes, tools, etc.), striations and manufacturing marks, trace evidence	Linking a cartridge with gun, marks with shoes, fibres with clothing
Locate and position things	Cell site analysis, shoe marks, pollen and plant traces, soil	Shoe marks at point of entry, tyre tracks in mud

(Continued)

Table 1. Continued

Investigative use	Technique/Examination	Examples
Timing events	Cell site analysis, combinations of evidence, e.g. shoe marks in blood	Mobile phone used near robbery scene, finger-marks in blood
Associating people with objects	DNA, fingerprints, trace evidence (fibres, glass, paint), firearms discharge residues	Links to guns, masks, weapons, clothing, drug packaging, a drinking glass, threatening letter
Associating events	Shoe marks, tool marks, fingerprints, DNA, mobile phone data	Linking crime scenes
Providing intelligence	DNA, fingerprints, shoe marks, tool marks, cartridge cases	National and local intelligence databases: DNA, fingerprints, shoes
Reconstructing events	Blood patterns, fire investigation, bullet recovery	Determining the seat of a fire, sequence of attacks using blood pattern analysis

This notion is more easily explained by images than in words, as can be seen in Figures 2 and 3. Both images show marks which can be used to identify the type of shoe that made them and potentially link this to the actual shoes that made the marks. The deposition of both marks can also be timed accurately or at least within certain limits. One is in sand on a beach and could only have been made between tides. The other is in concrete and could only have been made when this was wet. One mark is transient and will persist only for a few hours, the other is a permanent feature of the location. It is not the shoe mark in isolation that determines the significance of the evidence, but the mark together with the surrounding context.

We have considered the nature of crime and the processes involved in its investigation. The answers to questions raised by an investigation can often be provided by forensic science, and we have made initial linkages between certain questions and types of forensic evidence. Typically a combination of different types of

2. Shoe mark in sand

3. Shoe mark in wet concrete

evidence will be required and the significance of this will depend on the detailed context of the case. In the next chapter, beginning at the crime scene, we will explore how the police investigation and the scientific analyses are integrated to address the main issues in a case.

Chapter 3
Crime scene management and forensic investigation

> Q. Can you detect adrenaline that has been injected into someone who is now dead?
> A. I don't know – what is your hypothesis? Then we can work out how it can be tested.
> Q. Have you seen *Pulp Fiction*...?

Forensic science is driven by questions that arise outside the scientific laboratory in messy, distracting, and difficult situations such as crime scenes. The quote above illustrates how easy it is to fall into the trap of asking a question because there may be a scientific means of answering it rather than because it is the correct question to ask. Maintaining focus on relevant questions that can be feasibly answered by science requires the formulation of a realistic hypothesis about the events that have taken place. This is an important issue which we will explore in this chapter together with the principles and practice of crime scene management.

To ensure that the maximum potential evidence is obtained for an investigation, the process must begin at the crime scene. The purpose of crime scene management is to control, preserve, record, and recover evidence and intelligence from the scene of an incident in accordance with legal requirements and to appropriate professional and ethical standards. Although generally termed 'crime scene management', strictly speaking this should be referred to as 'incident scene management', as some scenes, such as an accidental gas explosion, are not crimes. In other instances, the central question of the investigation may be is this a crime? – therefore pre-judging the issue is not appropriate. For the sake of simplicity, we will refer to any scene as a crime scene whilst acknowledging this terminological inaccuracy.

In terms of their physical characteristics, scenes come in a huge variety but typical ones include houses, cars, and commercial premises. Less common examples of real scenes I have investigated are cabbage fields, secure mental institutions, railway carriages, fields, and motorways. Each type of scene presents different demands in how they are managed, but the most significant issue is why the scene is being examined. In many investigations there is more than one scene. In a homicide, in addition to where the body is found, the suspect's home and vehicle have the potential to be designated as crime scenes.

The person ultimately responsible for the scene is the investigating officer, who is invariably a police officer. In serious and major crime in the UK, the SIO works closely with a number of specialist personnel, particularly the Crime Scene Manager (CSM), who is responsible for advising on the detailed approach to deal with the scene. This includes the investigative potential of evidence types generally and specifically, the value of using experts in particular fields such as ballistics, blood patterns, or fire investigation, and the coordination of all of the events required to examine the scene effectively. This entails arranging and managing the post-mortem examination, agreeing the forensic strategy with the SIO, and maintaining ongoing communication with forensic science laboratories, individual experts, and the investigation team. Crime scene management in major crimes is physically and mentally demanding and requires high levels of knowledge of investigation and forensic science, excellent planning skills, good interpersonal communication, and team leadership. A CSM at a major incident will have a team of Crime Scene Investigators (CSIs) to carry out the necessary examinations.

Subject to the scene being safe to enter, the first stage of crime scene management is to secure the scene and ensure that it is preserved in the condition that is closest to the original state when the crime took place. This means clearing away any witnesses and bystanders who are present and making sure that no one else enters who does not have legitimate reason to do so. Physical security of the scene is managed by two sets of cordons. These are around the immediate (inner) scene (such as the house where a body has been found) and an area that is deemed relevant to the investigation and which may be much wider. This might include a street or part of a field. This outer cordon acts as a general control point in the early stages of the incident to ensure access is highly limited. A log is maintained, usually by a police officer at the inner scene, of all who enter, at what time, and for what purpose. This is a formal legal document and will be produced and may be inspected in any subsequent trial. Crime scenes attract a great deal of interest from those who have no right to be there, including busybodies, the press, and on occasions the offender. These basic security precautions control access, maintain the integrity of the scene, and minimize disturbance, interference, and contamination.

There are many myths and misunderstandings about contamination, some of which I will mention here. The first is that all scenes are examined using the highest standard of anti-contamination precautions (suits, overshoes, mob caps, gloves, etc.), which is not the case. Most volume crime scenes (burglaries, stolen vehicles) are examined by CSIs who are not wearing such protection, although they will generally wear gloves and masks when taking samples such as DNA. Secondly, the belief that contamination can be completely prevented by wearing the kinds of protection described above and by controlling a scene is unfounded. If you accept Locard's principle, then you have to accept that any examination of a scene is likely to disturb it and to 'contaminate' it in some way. Finally, the assumption that because someone has failed (for whatever reason) to follow 'recommended' operating procedures with regard to contamination does not mean that contamination will necessarily result and have an impact. We will touch on this issue in Chapter 7. Table 2 outlines the main principles and practices required for the management of most scenes.

Before commencing the examination of the scene, it must be accurately recorded in as near a state to the original as possible. This process is essential for a number of reasons. Firstly, it is a legal requirement (or at least an expectation) that a detailed record is available to the prosecution and the defence in any subsequent trial. It also acts as a repository of information for briefing those involved in the investigation who were unable to enter the crime scene. Finally, it aids reconstruction of events and allows interpretations to be checked and evaluated against the original facts. The scene is recorded using documents (notes, plans, diagrams, witness statements), images (stills, video and other specialist means), and sometimes audio notes also. All of these records serve to establish a contemporaneous record of the incident at the time of discovery and initial investigation.

Recovery of the evidence then begins in a planned sequence using an agreed search strategy. Since there are likely to be several people involved, it is essential that they each understand their particular role and that the entire scene is covered completely and systematically. What is recovered (and what is being searched for) will be related to the nature of the incident. Some of these things will be obvious – in a homicide by stabbing, finding the weapon will be high on the priority list, as would be bullets in a shooting. Documentation of various kinds is often more generally relevant to a criminal investigation. For example, a passport – is it genuine or false? Bank statements –

Table 2. Principles and practices of crime scene management

Principle	Practice	Comments
Control	Set up inner and outer cordons, crime scene log: timed and dated record of all who enter and exit, prevent all unnecessary access.	Use natural and manmade features (streams, roads, footpaths) to act as expedient boundaries, include rendezvous point for personnel attending scene, take account of press interest and needs of local residents.
Preserve	Minimize disturbance and contamination, limit access to essential personnel, common approach path, personal protective clothing: overall suit, hood up, mask, overshoes.	Outdoor scenes may need immediate protection from the elements or to be investigated immediately. Tents are usually placed over bodies to protect the scene and preserve confidentiality.
Record	Records must be contemporaneous, i.e. made at the time of the examination or shortly afterwards while the memory is still 'fresh'. If not, you may not be allowed to refer to them in any subsequent court proceedings or their accuracy may be challenged.	Includes written notes, sketches and plans, dictated notes, video, and still images. Important conversations or decisions should be recorded, dated, and timed.
Recover	Systematic recovery plan, logical sequence of recovery, suitably packaged and labelled items, detailed log of items recovered.	Allocate roles and tasks to particular individuals, monitor and log completion, take account of fatigue and potential overexposure in traumatic incidents.

is the individual deeply in debt or inexplicably rich? Finally, there is the possibility that evidence of criminality may also be found, such as drugs which may be irrelevant but could be central in a turf war between suppliers.

It is essential that all items recovered are labelled adequately to maintain continuity (from crime scene to court) and suitably packaged to prevent contamination, minimize damage, and maximize the potential of recovering evidence. The packaging needs of different types of items varies hugely, and it is one of the main responsibilities of the CSM that this conforms to laboratory requirements. Packaging a wet, bloodstained item in the wrong way could impact adversely on the prospects of DNA profiling, and failing to seal an item on which trace evidence could be found may result in it being excluded as evidence in a court hearing. The great variety in packaging is a consequence of the wide range of items that are encountered in investigations and the types of forensic tests available. Although the details are very different, each type of packaging follows a number of relevant principles which include:

- protecting the item from the outside world by acting as a barrier to transfer of adventitious materials;
- preventing loss of material (especially trace evidence) from the item;
- protecting those handling and transporting the item from risk of injury, e.g. from broken glass or sharp weapons;
- protecting individuals from exposure to infection such as blood-borne viruses (HIV, Hepatitis B and C);
- allowing easy transport, handling, and storage.

Table 3 provides a summary of the different packaging methods used for a wide variety of items and evidence types. It gives some indication of the level of detailed knowledge required to ensure each item is dealt with effectively and reaches the laboratory in a state that maximizes the potential evidence that may be recovered.

Table 5. Packaging of forensic evidence

Item	Packaging	Comments
Dry clothing	Paper bags	Store in a cool, dry environment. Also suitable for other dry materials (plant fragments, particulate debris, soil).
Clothing (wet)	Polythene bag	Dry as soon as possible in controlled conditions or transport immediately to laboratory.
Shoes	Paper bags	Shoes in polythene bags will go mouldy very quickly.
Documents	In folded card inside envelope or polythene bag.	Card prevents indented impressions being left on items accidentally.
Drinking glasses, footwear casts	Rigid cardboard box in polythene bag	Item must be secured inside box.
Cigarette ends	Sterile container, envelope, polythene bag, paper bag	Store in a cool, dry environment.
Hairs	Polythene bag or envelope	Store in a cool, dry environment.
Blood scrapings	Small sterile tube	Store in a cool, dry environment.

(*Continued*)

Table 3. Continued

Item	Packaging	Notes
Weapons/tools	Rigid transparent plastic tubes or cardboard box in polythene bag	Referred to as 'weapon tubes'. If in a box, items need to be secured in place.
Hypodermic syringes	'Weapons tube'	Rigid packaging protects those handling the item from 'needle stick' injuries.
Wet plant material	Polythene bags	Dry within 24 hours to prevent mould growth.
Broken glass	Polythene bag inside a cardboard box	
Paint fragments	In paper inside a polythene bag or envelope	Also suitable for small amounts of dry particulate material, e.g. powders.
Fibre tapings (lifts)	Sealed in individual polythene bags or envelopes	Polythene bags are better as the tapes can be viewed through the bag without being opened.
Wet particulate debris	Plastic Petri dish inside a polythene bag.	
Fire debris samples	Nylon bags or sealed glass/metal container, with foil lid if a liquid sample	Volatile liquids will pass through polythene.
Drugs	Polythene bags	Bags usually bear a unique serial number as a reference for each individual sample.

Although many aspects of scene management are procedural, it is essential at all times to consider the investigative implications of actions and decisions. The rational but contingent selection of one or more investigative opportunities may preclude others that are subsequently found to be of value. Such decisions need to be agreed and recorded as it may be necessary to justify them in subsequent trials.

Forensic strategies

In *Pulp Fiction*, there is a scene in which John Travolta plunges an adrenaline-filled syringe into Uma Thurman's heart which instantly revives her. There is a degree of medical authenticity to use of adrenaline in this manner, but it is far from routine or free from danger. A problem is that such fictional events can gain currency in the real world; something that is commonly referred to as the 'CSI effect'. In relation to the investigation of drug-related death, I was asked by the SIO if adrenaline could be detected in the body. No other information was given. Adrenaline is a natural substance that one would expect to find in the body, and unless there were huge amounts present it seemed to me unlikely that this was the most efficient line of inquiry. Such questions are instantly recognizable to an experienced forensic scientist as a distraction from the real issues, since they are framed in the wrong manner: as scientific questions as opposed to investigative ones. I replied that I didn't know (although I could have guessed) but also asked what he was trying to prove – this is how the question should have been framed. Only then was the *Pulp Fiction* scenario revealed. This approach is unfortunately quite common in police use of forensic science, although I am still unclear why that is the case. A much better approach is to formulate a hypothesis and identify scientific tests that might support or refute this. In the '*Pulp Fiction*' case, the hypothesis is that a syringe containing adrenaline has been plunged into someone's heart. How might we test this? Potential lines of inquiry might include:

- establishing if there were any witnesses (there were none, everyone was in a drugged stupor);
- looking for physical evidence of the event – syringes, ampoules of adrenaline (none were found as far as I was aware);
- testing samples from the body (if this is possible);
- examining the body to see if there is a needle-sized hole in the chest wall.

There are other possibilities, but of those suggested the examination of the body is quickest, simplest, most easily interpreted, and can be demonstrated to a jury or anyone else by means of a photograph. It gets my vote.

An SIO managing an investigation is surrounded by specialists and experts, police and otherwise. By definition, the SIO cannot know all the potential solutions and procedures, otherwise why would they need the other experts? Their primary role is to lead, coordinate, and focus the inquiry on the key investigative issues that must be resolved, and solve these by the best, quickest, and most efficient means possible. It is all too easy for imagination and pet theories to cause 'mission creep', but this can be avoided by using a small team to develop a forensic strategy which ensures constant focus on investigative issues and monitors progress regularly (daily in the early stages of a major investigation). A typical forensic management team consists of the SIO, deputy SIO, and all relevant forensic specialists: pathologist, fire investigator, shoe mark expert – whoever can help resolve the issues. The meetings are formally documented so that there is an ongoing record and everyone is clear what the priorities are. One of the main roles of the forensic management team is to manage the submission of items to the forensic science laboratory clearly specifying the needs of the investigation and constantly monitoring the results. From reviewing the submitted paperwork and attending any necessary meetings, the team of scientists in the case should be absolutely clear what their priorities are and be in a position to respond appropriately.

The forensic management team also helps address another common problem. Many scientists do not fully understand the investigative methods used by the police, nor are they kept fully aware of the changing investigative priorities in an inquiry. Table 4 gives some indication of the complexity of this problem from the range of evidence types, and therefore disciplines and scientists who may be involved in a single case. As can be seen, although the investigation of homicide typically involves examination of body fluids and weapons, it can also involve almost any other type of evidence depending on the nature of the case and the issues that need to be investigated, proved, or eliminated. Sexual offences, assault, robbery, and investigation of vehicle crashes can also involve a wide range of different evidence types. DNA tends to be of most value in offences against the person, and fingerprints can be useful in almost any investigation. Some crimes, such as fraud, use a comparatively narrow range of evidence types (handwriting, documents, computer examination) and others, such as drink driving, rely on a single type of evidence (quantification of alcohol).

It is sometimes more effective, quicker, or more convenient to bring the specialist to the crime scene. This is routine for crash investigation and fire investigation when the bulk of the examination takes place at the scene and is supported by follow-up laboratory work, but this is not typical of most other areas of forensic science. In both these areas, specialists are likely to be interested in events immediately prior to the incident and may also require information from eyewitnesses more readily available at the scene. The decision to call a specialist to a scene has to strike a balance between having another person at the scene who needs to be factored into the overall planning and the benefits of face-to-face discussions between the investigator and the expert. Shootings involving multiple deaths and fires

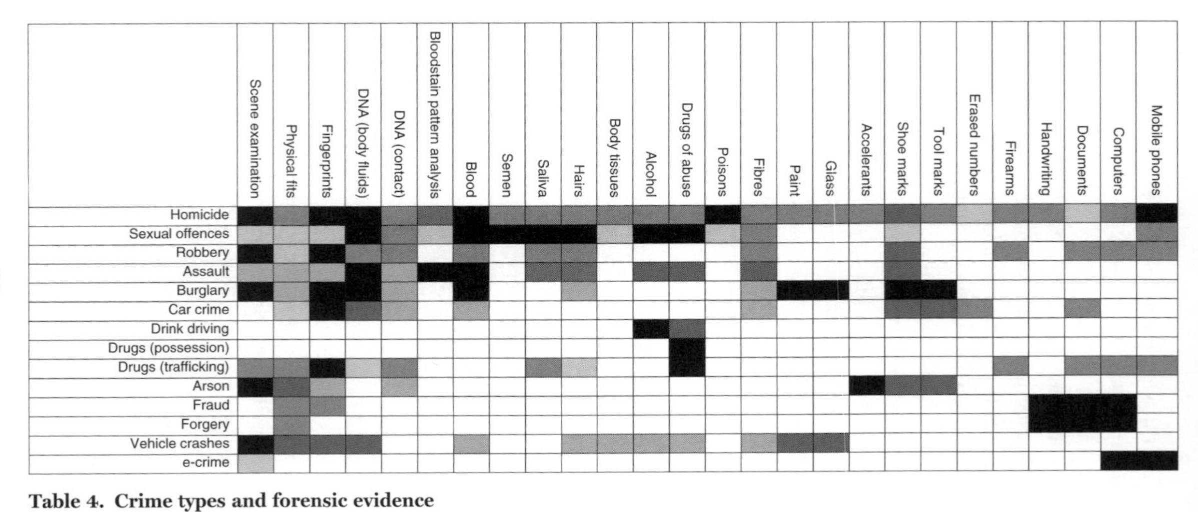

Table 4. Crime types and forensic evidence

Key: Black, regular or routine; dark grey, occasionally or where relevant; light grey, rarely; blank, not applicable

involving loss of life or high-value property are likely to involve specialists at the scene. In the former, a pathologist and firearms expert working together can combine information derived from body positions, injuries, locations of spent cartridge cases, and bullet damage to surroundings, to reconstruct the incident. The main drawback of this approach is that the bodies must be left in position until the specialists complete their examination, which can present difficulties if the examination is protracted or the location is outdoors.

An area of forensic biology which can be useful at the scene (and in the laboratory) is bloodstain pattern analysis (BPA). Information from this type of evidence can aid reconstruction of incidents, eliminate alternative explanations, and provide information about the sequence of events prior to interviewing suspects. Blood is a complex liquid that is a suspension of cells, proteins, salts, and enzymes. Blood droplets obey physical laws, and an understanding of these laws and how bloodstain patterns are created can allow scientists to interpret the crime scene. BPA requires an understanding of blood droplet dynamics and the expertise to link related blood patterns logically in light of other available information.

When force is applied to liquid blood, such as when a bleeding person is punched or struck with a weapon, the blood is dispersed as small droplets. The number and size of the droplets is related to various factors, including the force involved and the amount of blood present. Droplets can travel up to 4 metres from the point of impact and the distance travelled is related to their size: large droplets travel further than small ones. On landing, the droplets form a stain whose shape indicates the direction of travel and angle of impact. By triangulation using a number of droplets, it is possible to establish where the pattern originated from, and by inference, the location of the individual when the blow was struck. The angle of impact of each stain can be calculated by using the ratio of the width

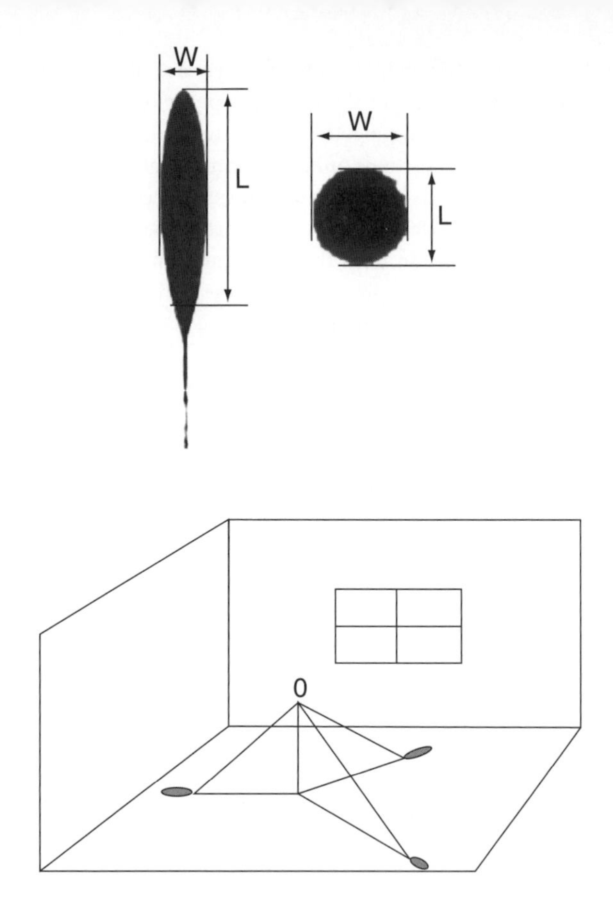

4. Calculating the angle of impact of bloodstains (top) and estimating the area of origin of blood spatter (bottom). The bloodstains above struck the surface at approximately 13° (top left) and 90° (top right)

Table 5. Characteristic bloodstain patterns. Interpreting these patterns in light of other information or evidence available can assist in reconstructing events at the crime scene

Smear/Contact	Indicates contact between a bloodstained item and another surface.
Drip	Free-falling blood droplets. In a linear pattern usually indicate movement (a trail).
Impact	A radial pattern of predominantly minute stains (around 1 millimetre) caused by force dispersing the blood. Generally correlated with violence.
Arterial	A distinctive pattern caused by blood escaping from a damaged artery,
Cast off (in-line)	A linear pattern of stains caused by blood being thrown from a moving object, typically a weapon such as a hammer in crimes of violence.

of the stain to its length, giving the sine of the angle. How to calculate the angle of impact and estimate the area of origin of bloodstains is illustrated in Figure 4. There are also a number of recognizable blood stain patterns which arise from specific events and can readily be identified. These are outlined in Table 5.

Not all patterns are readily identifiable, and some patterns are fragmentary and difficult to interpret. The best patterns are usually found on smooth surfaces on stable or immovable light-coloured objects such as walls or heavy furniture. Items that can be moved or disturbed may not be in their original position and this may affect the interpretation. The analyst must think in three dimensions and try to establish logical links between patterns. At the same time, any contrary evidence which arises must be considered as this may indicate a better alternative explanation. The type of information that can be provided by BPA includes: the number and location of attack sites; sequence of attack and where it began; the positions of the victim and attacker during the incident; how close an individual was to the blood spatter; if the body has been moved following the attack. The outcome of such an examination is subjective and rarely conclusive, but can still be extremely useful in certain types of offences such as homicides which involve blunt trauma and considerable bloodshed.

Good crime scene management is critical to the effective application of forensic science in criminal investigations. Failure to protect the crime scene, and to manage contamination and continuity of evidence, are likely to have major implications for an investigation and could preclude significant lines of inquiry. Identification of the relevant questions to be addressed and the formulation of specific hypotheses for scientific testing are essential. Leadership, communication, and teamwork are also central to ensure the right experts are addressing the relevant questions, at the scene and in the laboratory.

Chapter 4
Laboratory examination: search, recovery, and analysis

We now move from the crime scene to the laboratory and the various stages of recovering, documenting, and analysing evidence. Some of the principles and processes will now be familiar to us as they reflect those applied at crime scenes. The new dimension is the specific application of scientific testing of case items and the range of scientific disciplines involved. In this chapter, we will cover the types of examinations carried out in particular case types and the specific scientific and legal procedures required to meet the standards of criminal law.

Recovery of evidence

The modern forensic science laboratory contains a bewildering array of science and technology that is focused on providing answers and indicating valuable lines of inquiry in criminal cases. In addition to meeting formal scientific standards, it has to comply with legislation and legal procedures to ensure that the criminal justice system is best served. This chapter outlines how items are examined, the techniques that are used to recover evidence, and the potential range of methods available for analysis. The importance of minimizing contamination, maintaining continuity (chain of custody), and quality assurance are also explained. There is no standard structure that all laboratories follow as this will depend on the type and amount of examinations they carry out.

Some labs employ only a small number of people (say 15), others can be very large, containing many hundreds of staff. Some labs are part of police organizations, others are independent public sector or state organizations, and some are private commercial enterprises. In England and Wales, all the major forensic science laboratories are private companies. The privatization of these laboratories remains a matter of considerable debate in terms of whether this is in the interest of justice. Detractors of this arrangement suggest that it will inevitably lead to erosion of standards and a focus on profits to the detriment of justice. Supporters point to other private industries with very high technical standards, such as the airline industry, as evidence that private enterprise does not mean lower standards of science and technology. Only time will tell how effective these arrangements are. In most other countries around the world, including the USA and Australia, commercialization of forensic science is rare and tends to be in the form of small companies which provide one specialist type of evidence such as DNA profiling.

Whatever their funding basis, most medium-sized labs (say around 100 staff) will have broadly similar structures based on the types of investigation they are involved in and the scientific disciplines that they need to do this work. Very few labs in the world, perhaps none, have the capability to carry out every forensic examination, and most seek to balance the skills they have with their users' needs. Table 6 provides an overview of the scientific disciplines, departments, and case types found in a medium-sized laboratory. In terms of case numbers, the bulk of the work will be volume crime (burglary, car crime) and drugs analysis, where small numbers of items will be examined in individual cases. Although serious offences will be small in number, the workload in these cases is likely to involve many more examinations, sometimes of hundreds of items.

Irrespective of how the laboratory is structured, the first stage of any examination is the recovery of materials. In small laboratories,

Table 6. Disciplines, departments, and case types in a typical forensic science laboratory

Section/Evidence type	Examinations	Comments
Evidence recovery unit	Routine initial examination of a wide range of items such as clothing, weapons, etc. to recover evidence for further analysis using standard techniques: visual examinations, taping, and sweeping.	This is the first stage in most examinations and is carried out by an assistant under the supervision of a case-reporting scientist.
General chemistry	Analysis, comparison, and identification of a wide range of materials and chemical substances such as greases, waxes, plastics.	Involves a wide range of cases, such as burglary, thefts of material, major crime, and road traffic crashes.
Chemistry trace evidence	Analysis, comparison, and identification of minute quantities of paint, glass, soil, and other chemical traces.	Involves cases similar to those above. In a small laboratory, may be part of General chemistry.
Marks	Comparison of shoe marks, tool marks, tyre marks, and manufacturing marks of various kinds. Will also include the provision of intelligence and linkage of scenes.	Widely used in volume crime, especially burglary, and major crime such as homicide.

(Continued)

Table 6. Continued

Section/Evidence type	Examinations	Comments
Drugs	Analysis and identification of drugs of abuse including synthetic and natural products and prescribed drugs.	Includes seizures from individuals as well as bulk importations in trafficking cases and materials from clandestine laboratories.
Toxicology	Identification and quantification of alcohol, drugs, and poisons in body samples in suspicious or sudden deaths and drink driving cases.	Poisoning is rare, but toxicology screening is a routine part of homicide investigation (and other serious cases) as drug use may have a bearing on the behaviour of victims and suspects.
Biology	Routine examination and comparison of a wide range of biological evidence (e.g. blood, semen, saliva) not dealt with in specialist sections.	Cases are typically violent or sexual assaults and homicide.
DNA	Genetic analysis of biological fluids, tissues, and stains by a range of techniques.	Includes examinations of mixtures of body fluids and paternity/maternity testing.
Fibres and hairs	Comparison and identification of natural and synthetic textiles, human hair, and animal hair.	Typically confined to serious cases such as sexual assaults and homicide, but increasingly used in wildlife crime.

(Continued)

Table 6. Continued

Section/Evidence type	Examinations	Comments
Botanical evidence	Examination and identification of a wide range of infrequently encountered materials such as plant fragments, seeds, wood, pollen.	In many cases, external experts will be consulted due to the specialist skills involved.
Documents	Examination of questioned documents in fraud and counterfeit cases – contracts, wills, letters, passports, currency - to establish ownership or authenticity.	Includes examination of documents from computer printers and faxes, their method of production and the analysis of ink.
Handwriting	Examination of handwriting to attribute or eliminate a putative author or connect documents that may have been written by the same author.	Involved in a wide range of cases including fraud, robbery, homicide.
Fire investigation	Examination of fire scenes and analysis of debris from fires or flammable liquids ('accelerants').	A great deal of this work is done at scenes, with the main laboratory work being the identification of flammable liquids.
Firearms	Examination and test firing of pistols, rifles, military weapons, and related devices. Identification of firearms discharge residues (FDR).	Examination of firearms and the identification of FDR is usually done in separate sections due to the potential contamination issues.

(Continued)

Table 6. Continued

Section/Evidence type	Examinations	Comments
Fingerprints	Comparison, identification, and enhancement of finger marks. In most cases, the laboratory work is confined to enhancement of marks, with comparison being carried out in a separate fingerprint department.	Many laboratories do not have a fingerprint section but most will have some capacity to enhance or visualize marks for examination by fingerprint experts.
Crash investigation	Investigation and reconstruction of road traffic incidents. Much of this work is done at scenes, but there may be follow-up analysis required.	Examination of tachographs and damaged vehicle parts, calculation of speeds and trajectories of vehicles to reconstruct crashes.
Digital evidence	Examination of computers, networks, and mobile devices (phones, PDAs, SATNAV, etc.).	Probably the most rapidly expanding area of forensic science. Digital devices are now so widespread that they are involved in many different investigations.
Photography and imaging	Routine record photography and analysis of imaging devices such as media from CCTV and still cameras.	Extensively used in a wide range of investigations, for example to identify individuals at or near crime scenes and for presentation of evidence in court.

individual scientists, perhaps working with an assistant, will do this. In larger laboratories, there will be an evidence recovery unit staffed with individuals trained to recover all of the potential evidence from items using many recovery techniques. Before the examination commences there are some basics steps to take. The first is to ensure that you have all the available information to carry out an examination, including relevant witness statements and police reports. This usually means a phone call to the investigating officer to check any facts which may have changed (this can happen overnight in major cases). Secondly, some basic planning needs to take place to ensure you have identified an appropriate sequence of examinations (you can't look for the red fibres from the jumper of the victim until you know what these look like).

This of course raises an issue of potential contamination, so you need to make sure that all relevant examinations are separated in space (different benches in different labs) and time (different days) and wearing different protective clothing. If there are three suspects, two victims, and a scene involved in a case, this will take some careful thought and planning. The risks of contamination are higher given that all of the materials from the incident are now in one place (the laboratory) and will probably be examined by a single scientist. However, the items are now in controlled conditions and can be managed more easily and effectively than at a crime scene. From the outset, systematic, stringent procedures are taken to prevent contamination and records that demonstrate compliance with these procedures are made and retained. What constitutes contamination and the steps taken to avoid it varies in the different disciplines of forensic science, and to an extent in different laboratories. Trace evidence – glass, paint, soil, hairs, fibres, and other particulate materials – are particularly prone to contamination, and the following steps are commonly used to minimize contamination:

- Items from different sources, e.g. the scene, suspect, and victim, are stored from the outset of the examination in separate places.

- The sequence of the examination should minimize the risk. Where possible the trace evidence is recovered before the control sample (the potential source) is examined. Once the traces are recovered, the opportunities for contamination are considerably lower.
- Items from which trace evidence is to be recovered are examined in different locations and at different times, generally a minimum of a working day apart.
- Different lab coats, examination benches, and instruments are used for each related set of items, e.g. a set of clothing from one individual. Extensive use is made of disposable instruments and protective clothing.
- The instruments assigned to a search bench are dedicated to it and do not leave it. The lab coat used is stored there until the case is finished.
- All of the above details are noted in the case file.

Examination of items

In this section, we will deal mainly with biological evidence, since aspects of chemistry (trace evidence and drugs analysis) are covered in later chapters. The examination of clothing for body fluids, such as blood, and trace evidence would commence by choosing an appropriate bench which will be cleaned, disinfected, and protected by a layer of clean paper. The examiner will wear a freshly laundered lab coat, new gloves, face mask, and disposable cap. The tools used in the examination (pens, forceps, etc.) will be located at that bench and will not be removed from there. The examination of an individual item proceeds as follows:

1) The label of the item is checked and compared with the relevant paperwork. There should be no significant discrepancies. If there are, this will have to be explored to eliminate any problems with the integrity or continuity of the item. The item number (or description in some jurisdictions) acts as a unique identifier and will be referred to in reports and in court so must be recorded exactly in the case notes.

2) The integrity of the packaging is reviewed. The item should be sealed and the packaging intact. Any deficiencies, such as damage or poor seals, must be noted. Where there is a significant problem, such as an unsealed item or breached packaging, the item may not be examined. Detailed notes of the packaging and sealing are recorded in the case notes.

3) The package is then opened in a different part from the original seals (which must be kept intact) and gently the item is laid out on the bench. The item is then briefly scanned for visible material of interest that might be easily dislodged and lost. This should be removed and retained in a separate labelled package (typically a small polythene bag). Minimal handling should be used at this stage.

4) If required, the surface of the item is then taped using transparent adhesive tape to systematically and completely recover extraneous trace evidence. Tapes from each item are packaged separately and labelled with the item details (description or number) and the location where the tape was taken from (e.g. front right sleeve, left rear, etc.).

5) Alternatively, when recovering particulate evidence such as glass and paint (and when there is no need to recover fibres), the item can be brushed to remove any microscopic particles after a careful and thorough visual and low-power microscopic examination.

6) Following recovery of the trace evidence, the item is examined visually, slowly and systematically, for other relevant evidence. Specialist lighting, e.g. from fibre-optic lamps, is often used for this process.

7) The item is then described—a shoe, jumper, or knife—in sufficient detail for it to be identified readily in future, e.g. in court. Its condition (old, new, worn), and any other significant or distinguishing features such as stains, marks, or damage, should also be noted. In cases involving fibres, the composition of the garment given on the garment label is noted.

8) In cases where items subsequently may be searched for fibres matching the item being examined, a control sample should be removed. This must be representative of the item and include all fibre types and colours.

9) The case notes should reflect what the item was examined for, its description, and any findings and interpretations, accurately and concisely, supported by measurements, diagrams, and photographs where relevant.

The purpose of examining items in this manner is to ensure that all evidence is recovered; to identify any relevant materials, such as body fluids, present; to produce accurate, detailed notes about the nature of the item; and to determine which analyses will be carried out next (such as the individual stain for further attention). These notes will be continually updated throughout the examination of the case and act as a detailed history of events and information received as well as analytical results. For example, if a case briefing is held, a record of the meeting will be stored in the case file. The case file will subsequently be used as the basis for drafting reports and to act as an *aide-memoire* when giving evidence in court. It should also be a transparent record of how the item was examined, by whom, for what purpose, at what time, and in what sequence. All of these matters are of potential interest during a trial.

Blood and body fluids

In physical and sexual assaults, body fluids can be shed and transferred to clothing, objects, and weapons. In addition to identifying the source, that is the individual from whom the stain has come, the location, amount, and pattern of staining can be important in interpreting findings. A common explanation for bloodstaining found on the clothing of individuals accused of assault is that they were near the attack but did not take part in it. The staining therefore must be examined to determine if there is any evidence to support or refute this statement. Saliva staining inside a mask can indicate that it has been worn and by whom. Semen and saliva are also routinely encountered in sexual offences on body swabs (e.g. from the vagina or mouth) and clothing, particularly underwear. Locating and identifying blood, semen, saliva, and other biological materials form a routine aspect of forensic biology.

The first stage of the examination uses a number of simple tests (so-called presumptive tests) that can give an initial indication of the type of stain which is then further analysed by confirmatory tests. Dried bloodstains have a characteristic red-brown appearance and are usually readily recognizable. A number of presumptive tests can be used to indicate the presence of blood, all of which rely on the catalytic activity of haemoglobin, the protein found in red blood cells. The Kastle Meyer (KM) test is commonly used to identify bloodstains. This is based on the oxidation by peroxidase of the colourless form of phenolphthalein to give a bright pink colour. For the test to be reliable, the distinctive pink colour must appear almost instantly, since the phenolphthalein will gradually oxidize and turn pink in the air anyway. The combination of the visual appearance of the bloodstain followed by a satisfactory KM test is generally regarded as sufficient to establish that a stain is blood. Stains that give a KM positive test but do not resemble blood may be mixtures of blood with another body fluid, such as saliva due to bleeding from the mouth. Alternatively, the stain may not contain blood, as a range of other biological materials and some chemical oxidants can give false positive reactions, although these are rare. The test is applied by taking a small piece of filter paper and rubbing it gently against the stain. Since DNA profiles can be successfully obtained from extremely small stains, one must ensure that the stain is not lost or destroyed in this process.

Semen consists of a fluid (seminal plasma) containing many millions of sperm as well as proteins, salts, sugars, and ions. Seminal staining on fabrics is usually whitish in colour but this can vary, especially if it is mixed with other body fluids. It can also form deposits that are colourless and can therefore be difficult to locate on some items. Establishing the presence of seminal staining relies on detecting seminal fluid and sperm. Seminal fluid contains a high concentration of the enzyme acid phosphatase that can be detected using a presumptive test known as the acid phosphatase (AP), or brentamine, test. This test relies on the formation of a purple azo dye from brentamine due to the catalytic action of acid phosphatase. The deeper the colour and the faster it appears (usually a few seconds), the more confident one can be that the reactions are due to seminal fluid. Other body fluids, in particular vaginal fluid, can also react to this test but the colour is different (more pinkish) and the reaction takes longer (over 30 seconds). However, given that most stains encountered in sexual offences are mixtures of semen and vaginal fluid, the difficulties in interpreting such a test will be evident. Following a strong AP test, the presence of sperm can identify semen. This is done by microscopic examination of a small amount of material extracted from the stain. Sperm have a characteristic appearance (see Figure 5), consisting of a head and tail section, and can be stained using histological dyes. The amount of semen found on vaginal swabs can be used to estimate time since intercourse, although this method is fairly crude.

Saliva is encountered in sexual offences and a wide range of other cases, such as robberies (on masks and gags) and homicides. Saliva is secreted by the salivary glands and contains water,

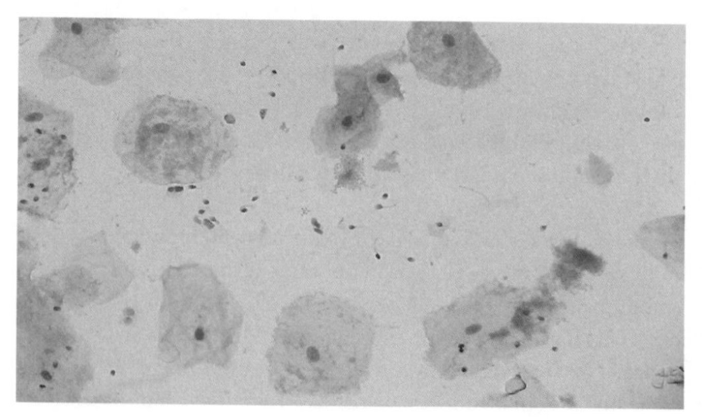

5. Sperm and vaginal cells stained with haematoxylin and eosin

mucus, proteins, salts, and enzymes. One of the enzymes present – amylase – is found in very high concentrations in saliva compared to other body fluids and its detection can indicate the presence of saliva. Saliva deposits usually form colourless stains, therefore identification must take into account the circumstances and location of the stain. The presence of epithelial cells typical of those from the mouth can sometimes be used as a confirmatory test, but these cells are similar to those found in the vagina (and elsewhere in the body) and are therefore of limited value.

The tests described above usually constitute the first steps in cases involving biological evidence, but the scientific procedures in a major case will require extensive examinations by other scientific disciplines with the aim of establishing answers to

Table 7. Samples and examinations from a male suspect in a sexual homicide

Sample	Purpose/Analysis
Buccal (mouth) swab	DNA reference sample
Blood	Therapeutic drugs, drugs of abuse
Blood	Alcohol analysis
Urine	Alcohol, drugs
Genital/Anal swabs	Body fluids – blood, saliva, vaginal material
Fingernails	Trace evidence, e.g. fibres, hairs, DNA from body fluids
Head hair	Hair reference sample
Pubic hair	Hair reference sample
Underwear	Trace evidence, body fluids, contact DNA
Clothing	Blood, body fluids, trace evidence

investigative inquiries. Table 7 gives an indication of the typical samples submitted to the laboratory from a suspect in a sexual homicide and the purpose of their examination or analysis. Blood samples of different types are required depending on the purpose of the examination. For example, the DNA reference sample contains an anti-clotting agent which enables extraction of DNA, and the sample for alcohol analysis contains a preservative that prevents infection by bacteria that can produce or metabolize alcohol and therefore cause misleading results. Urine samples are also required for alcohol and toxicology analysis, since alcohol and metabolites from drug breakdown can be detected in urine even when none is present in the blood. Depending on the detailed nature of the case, swabs from parts of the body including the genitalia and anus will be taken from the victim (by forensic medical examiners) to be examined for body fluids. Head hair and pubic hair samples are used for reference purposes should any hairs be found on the victim that require microscopic comparison, although this is increasingly rare given that DNA profiles can be obtained from hairs. Combed head and pubic hair samples may also be examined for fibre transfer. Finally, clothing known to be worn at the time would be submitted for examination for any evidence of relevance.

Table 8 provides a summary of the types of examination carried out in laboratories and the range of analytical techniques involved. Although many specialist techniques are used, we can say that in almost all instances the process begins with a visual examination of the items involved, usually to recover the evidence. The most commonly used range of techniques is microscopy, of which there are a number of specialist types that have particular application for the examination and comparison of trace evidence. Presumptive testing is also widely used to screen biological and chemical substances. The main analytical method which underpins most forensic biology work is DNA profiling, since this can identify the donor of almost any type of biological fluids or tissue. A large number of analytical techniques is used to identify the diverse range of substances encountered in forensic chemistry.

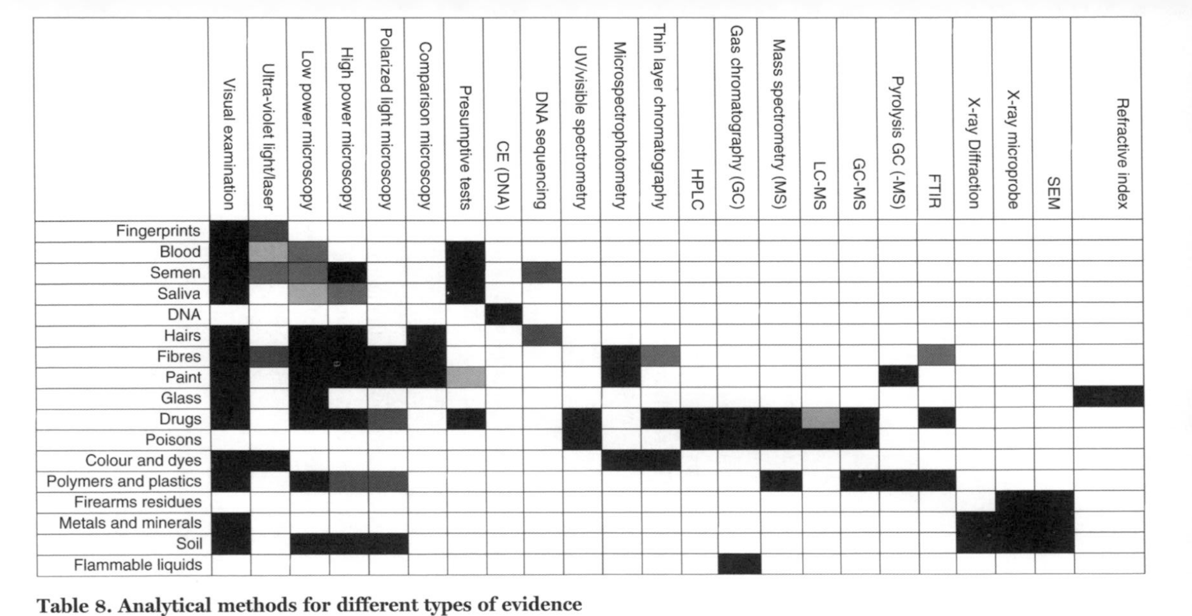

Table 8. Analytical methods for different types of evidence

Key: Black, regular or routine; dark grey, occasionally or where relevant; light grey, rarely; blank, not applicable
HPLC: high-performance liquid chromatography; CE: capillary gel electrophoresis; FTIR: Fourier transform infrared spectrocopy; SEM: scanning electron microscopy

The particular method used will depend on the nature of the substance involved, whether it is organic or inorganic, solid or liquid, or present in trace quantities or large amounts. A number of the techniques mentioned in Table 8 are further described in other chapters: Chapter 5 (DNA), Chapter 7 (trace evidence), and Chapter 8 (identification of drugs). It is not possible to cover all of the techniques in the table, but for those readers who wish to explore this area further, suitable references are provided.

Physical fits

This chapter is about the recovery and analysis of evidence, but on occasions both of these processes can be remarkably simple. A physical fit occurs when two separate items fit together in such a manner that it becomes instantly recognizable that they were originally one item. This rarely requires any analytical equipment other than occasionally a low-power microscope, nor does it require any scientific interpretation. For example, a 24-hour service station in London was robbed and the assistant was handed a McDonald's takeaway bag to put the cash in. As the robber took the bag, the assistant held on to it, gently tearing a small piece off. The fragment contained parts of the multi-coloured McDonald's logo and fitted perfectly with the torn section of the bag recovered from the robber's address. A photograph of the items fitting together as one was taken to accompany the short statement and to illustrate the nature and significance of the evidence. A similar example to this is illustrated in Figure 6. In a second example, fragments of glass from a broken container and a piece of burnt fabric were recovered from the scene of a fire and were suspected to be parts of a petrol bomb. The glass fragments pieced together to form most of a milk bottle and the fabric was a badly burnt fragment from a white vest which bore traces of a flammable liquid. The remains of a white vest were recovered from a suspect's premises and were compared with the burnt fragment. In this case, the match was less clear-cut and required careful microscopic examination of both the vest and burnt fragment to establish a fit.

Nonetheless, the burnt fragment fitted together with part of the left shoulder strap of the vest, showing that they were originally one item.

Until now, we have been considering the process of examination, the physical actions and logical steps by which it is achieved. There is a great deal more to an effective examination than following a process. In searching, there is a need to be alert to potential evidence that may be unforeseen. The examiner must take an intelligent, inquiring approach which is suitably detached and dispassionate. It is reasonable to have general expectations – we know that individuals involved in violence where there has been bloodshed may get blood on their clothing – but we should not be motivated by such expectations in individual cases to achieve any particular outcome. We proceed from observation to analysis and then by inference to interpretation. Statements and reports should also follow this pattern as we move from observation (fact), analysis (fact), interpretation (a mixture of fact and opinion), to conclusion (usually opinion). A report should separate fact and opinion so that this is clear to the reader. The discipline of this procedure is valuable; otherwise we can be prone to errors that arise from statements which are opinion but appear to be fact. In other words, the sequence is not only procedural but cognitive. Considerable care is needed to prevent the synthesis of fact and interpretation in our minds because we are so used to handling particular types of analyses and interpretations. For example we categorize a blood pattern as 'impact' on the basis of an inference from our observation of certain characteristics. This is nonetheless a contestable assertion and not a fact, an assertion we need to explain and justify if necessary. Much of the value of forensic science derives from the combination of scientific testing and rational transparent process which can be judged by others as objective (or not). It is the scientific approach that is objective, not the scientists: scientists are no more objective than anyone else.

This chapter has covered the fundamental processes of forensic examination: search, recovery, and analysis. It has also described

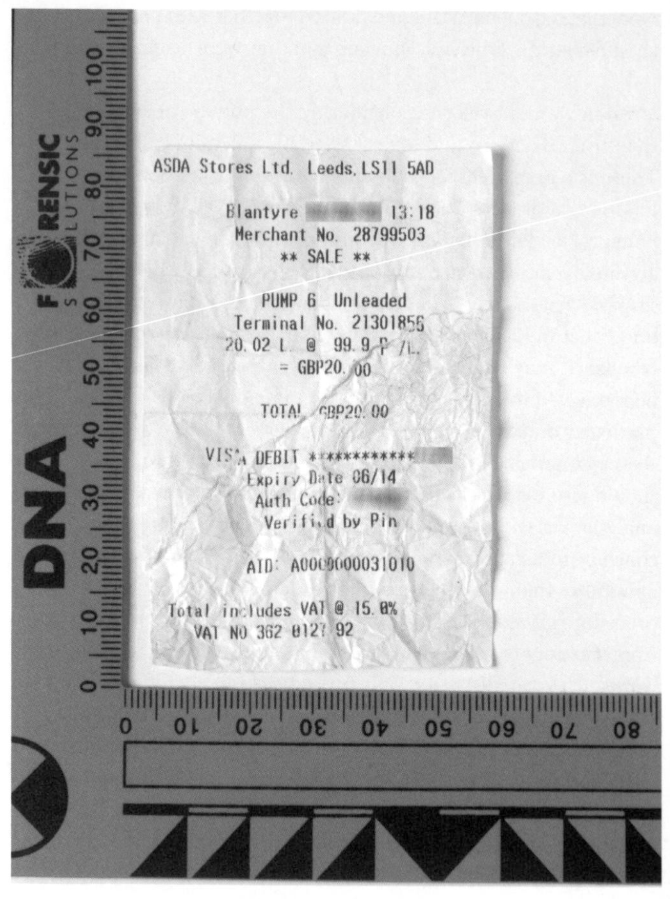

6. Physical fit of two parts of a petrol receipt

the range of methods involved in such analyses, how to prevent contamination and maintain the integrity of the evidence, and some of the issues that arise in the interpretation of evidence. In the following chapters, we will consider these issues and other specific areas of forensic science in more detail.

Chapter 5
DNA: identity, relationships, and databases

In not much more than 20 years, the law has moved from considering DNA profiling as usurping the role of the jury to fully embracing it as a distinctive and positive contribution to the investigation and prosecution of crime. DNA profiling has also set a new standard for forensic evidence, sometimes referred to as the 'gold standard'. It has followed a traditional path from discovery to application which is typical of new scientific developments, in that research evidence has been published, peer reviewed, and challenged in the scientific community. Ironically, many other areas of forensic evidence do not live up to the new standards set by DNA. In this chapter, we will look at the biological basis of DNA profiling and how DNA is analysed and interpreted in different case types. We will also consider the nature of identity, a central issue in forensic science of relevance to this chapter and subsequent chapters, and the use of databases in the investigation of crime. The discovery of DNA profiling by Sir Alec Jeffreys in the mid-1980s was the single most important breakthrough in the investigation of crime since the discovery of fingerprints and, amongst other things, led to the establishment in England of the world's first DNA database, in 1995. The impact of DNA profiling has been immense. This is due to the fact that it can eliminate or identify an individual from minute traces 'invisible' to the naked eye with great confidence. This chapter explains the structure of DNA, the mechanism of DNA profiling, how it is interpreted, and the operation of DNA databases.

What do we mean by identity?

Much of forensic science is aimed at identifying things – people, objects, substances – but what do we mean by 'identify'? Identity means different things to different people and has a commonsense usage (which varies depending on context) as well as philosophical interpretations. Identity is central to the criminal justice process since we must be certain that a person arrested or found guilty is unequivocally who we believe them to be (irrespective of who they say they are). So what do the terms 'identity' or 'identify' mean to a forensic scientist? You may be surprised to find that it depends whom you ask and that different disciplines within forensic science use different standards and criteria to identify things. Although the terms are not universally used, most forensic scientists will draw a distinction between classification of things (placing an object in a defined category) and identification (the recognition of uniqueness – that something is *one* of a kind). Classification can be a continuous process – a car, a red car, a red sports car, a red sports car with damaged bodywork – which increasingly moves towards narrower or smaller categories; whereas identification is a final and categoric determination of uniqueness – *the* red sports car with the damaged bodywork abandoned at the scene of the crash. There is only one such vehicle.

Some forensic scientists, especially in the USA, use another word which helps avoid some of this confusion – 'individualization'. This was coined by one of the pioneers of forensic science, Paul Kirk, and has the benefit of being unambiguous. If you individualize something, it is one of a kind, unique; it has not merely been classified, no matter how few things there are in the class. Kirk suggested that individualization is the primary aim of forensic science, but it will be clear from earlier chapters that this cannot be the full story. There are very many instances when individualization cannot take place for reasons of practicality or technological limitations, and there are many instances when individualization is not necessary to answer an investigative question. Many very experienced forensic scientists (Dave Barclay, Pierre Margot, and the late Stuart Kind) consider that answering investigative questions is the aim of forensic science, irrespective of how this is done, and this is a view I share. Evidence or intelligence that falls a long way short of individualization can be relevant and valuable in a criminal inquiry – a very poor, fragmentary finger mark could eliminate someone from an inquiry. What is more important is that the weight of the evidence is assessed in the context of the case, as we will see later. Identification and the nature of identity are live issues in forensic science at the moment and we will return to them in various chapters in this book.

DNA and the human genome

DNA (deoxyribonucleic acid) is the genetic material of most living organisms and plays a central role in determining hereditary characteristics. DNA is a major component of the chromosomes found in the nucleus of each cell in the body and is also found in cell organelles called mitochondria (mitochondrial DNA). DNA consists of two complementary chains of molecules wound around each other in the form of a double helix. Each chain consists of molecules of the sugar deoxyribose linked by phosphate molecules. Attached to each sugar is one of four nucleotides or bases: adenine (A), thymine (T), guanine (G), and cytosine (C). The relationship between pairs of these nucleotides forms the basis for the complementary chain, in that adenine only pairs with thymine and guanine only with cytosine. This means that when the chains separate for replication each can form a complementary version of the other, resulting in two identical molecules. The diagram (Figure 7) illustrates the structure of DNA and the relationship between DNA and other parts of the cell.

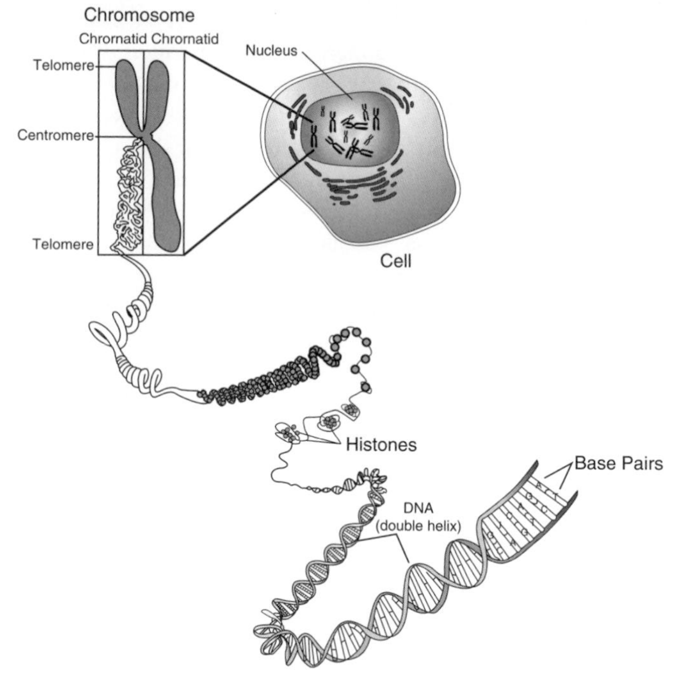

7. The structure of DNA, its relationship with chromosomes, and location within the cell

Human cells have 22 pairs of matched (or homologous) chromosomes and a pair of sex chromosomes (XX, female or XY, male) that make up the genome, the entire complement of genetic material. Each chromosome consists of a single continuous strand of DNA, together with proteins known as histones that support the organization and packaging of the DNA. Genes, the units of heredity, consist of two components (or alleles), one of which is inherited from each parent. Around 25% of nuclear DNA is involved in the expression and regulation of genes. The remainder of the genome does not appear to play a role in gene expression and contains what is referred to as non-coding DNA. Amongst other things, the non-coding parts of the genome contain large amounts of DNA that consist of repetitive sequences. This includes tandem repeat DNA, of which short tandem repeats (STRs) are the most important in forensic terms. We do not know how these STRs have come about. Although this DNA is non-coding, it still forms part of the allele and is therefore inherited in a predictable pattern. In STRs, the core repeat element is typically between 1 and 6 base pairs. Different alleles will have different numbers of repeats of the core sequence, and there are thousands of STRs scattered throughout the human genome. The most useful STRs consist of different numbers of tetranucleotides (sets of four nucleotides) that form each allele and occur in particular frequencies in human populations. It is from this detectable variation that the power of DNA profiling arises. Because we know these alleles are inherited independently from one another, their frequencies can be multiplied together to determine how common each combination of alleles (the DNA profile) is in the population.

Analysis of DNA

The first step in this process is the extraction and purification of the DNA. The method for this varies depending on the tissue or stain type involved, and some tissues, such as epithelial cells, are easier to deal with than others, such as bone. In this process, cell membranes are disrupted, proteins are denatured, and the DNA is separated from the denatured protein. A particular process of extraction is required for stains containing semen and semen mixed with other epithelial cells from body fluids such as vaginal fluid or saliva. This process is known as differential extraction and relies on the fact that sperm are resistant to the enzyme that breaks down cell membranes and require the addition of a reducing agent to break down the wall of the sperm. Following extraction, it is important to quantify the amount of DNA that has been recovered, as this can vary widely and the amount used in the next stage of the process is important. Most commercial kits require between 0.5 nanograms (10^{-9} grams) and 2.5 nanograms of DNA for optimum results. A range of methods can be used for quantification electrophoresis, ultraviolet spectroscopy, and fluorescence spectroscopy.

DNA can be amplified for analysis using the polymerase chain reaction (PCR) and quantified in 'real time'. The PCR process

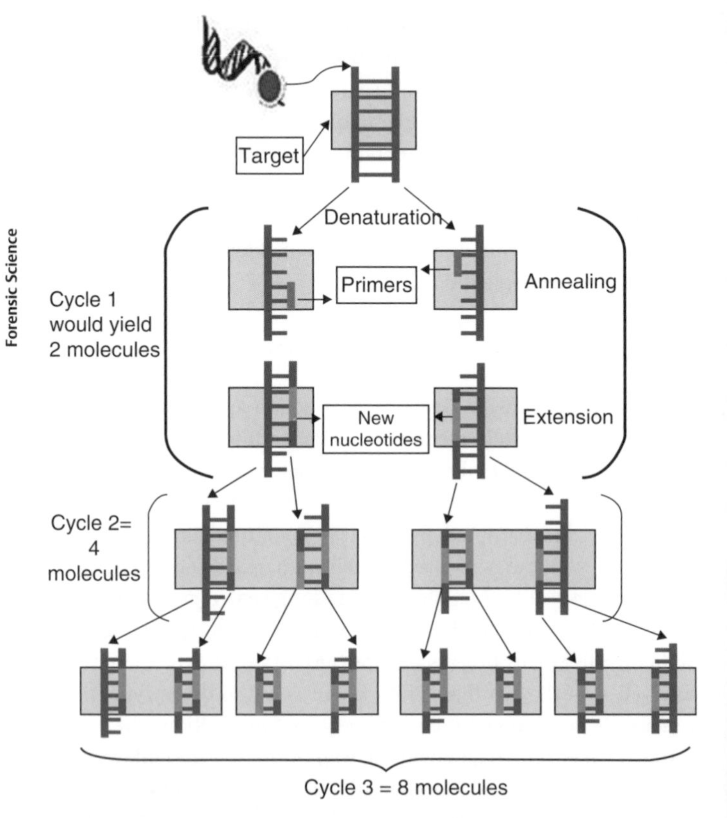

8. The polymerase chain reaction. At the end of each cycle of heating and cooling the amount of DNA present doubles

(Figure 8) results in an exponential increase in the amount of targeted DNA (i.e. the DNA in the alleles being analysed) being amplified and explains why profiles can be obtained from such tiny samples. Two short pieces of synthetic DNA known as primers are required to flank the target DNA to identify it. The design of these primers ensures that only human DNA and not that from other species is amplified.

There are around 20 commonly analysed STRs used for forensic purposes. These have been selected on the basis of important characteristics required in forensic profiling:

- The alleles are not linked to other genes such as those influencing physical characteristics or those associated with inherited disorders.
- The alleles are inherited independently of each other.
- The alleles are small and stable, so being comparatively resistant to degradation by heat, moisture, bacterial action, etc.
- They are highly discriminating due to the high level of 'variation' exhibited.

Analysis takes place in multiplexes, that is, many tests are carried out at once in the same test tube. There are two main systems used around the world that have been developed by commercial organizations. The AmpF/STR SGM*plus*® system (Applied Biosystems) is used in the UK and many other countries. In the USA, different systems are used that are based on the standard set of STRs used by the DNA database known as CODIS (Combined DNA Index System). CODIS uses a set of 13 STRs which are included in two commercial kits (AmpF/STR Identifiler from Applied Biosystems and PowerPlex 16 produced by the Promega Corporation). The number of amplification cycles (28 or 32) used depends on the particular kit and will produce around 100 million to 1 billion copies of the target DNA. It is important that this number is not exceeded as it can lower the quality of DNA produced and compromise the interpretation of the results. The process of amplification takes place in a thermocycler that carefully controls the heating and cooling cycles to ensure predictable, high-quality results. The particular advantage of PCR is that it can produce analysable quantities of DNA from only a few cells. A consequence of this is that great care needs to be taken throughout the entire process to ensure that the result obtained is a true one originating from the sample and not some external source of DNA, such as the analyst or the environment, due to contamination. Routine quality-assurance procedures are taken for this reason, including:

- regular cleaning and decontamination of laboratories and equipment;
- use of disposable equipment (such as pipette tips);
- wearing of protective clothing, such as masks and mob caps, by analysts at all stages in the examination of the item and analysis of the DNA;
- separate dedicated laboratories, equipment, and work streams for the handling of reference materials (i.e. those known to have come from suspects, witnesses, or victims) and crime samples;
- use of independently accredited methods of analysis and standard operating procedures that include positive and negative controls in each test run;
- maintenance of staff (police, CSIs, scientists) and supplier (e.g. sample kits) databases for investigation of contamination events.

The purpose of all of these steps is to minimize potential contamination, or maximize the likelihood of discovering contamination events in order to ensure they have no impact on the investigation and prosecution of the case. Such events inevitably occur, but the above procedures will reduce their impact.

Analysis and interpretation of DNA profiles

Different alleles in DNA profiles vary in molecular weight and can therefore be analysed by the technique of electrophoresis. This separates molecules on the basis of their electrical charge and mass. There are many different types of electrophoresis, but the most widely used method for DNA analysis is capillary gel electrophoresis. Coloured dyes are attached to the DNA and these are detected by a laser to produce an electropherogram (EPG) similar to that shown in Figures 9 and 10. The peaks in the EPG represent the alleles detected. The grey peaks are fragments of DNA of known size to allow calculation of the allele sizes in the profile. From left to right on the EPG, the DNA fragments become larger and therefore are more prone to degradation. Occasionally there are tiny peaks near the bigger peaks. These are due to an artefact called 'stutter' and can be safely ignored for our purposes. The first figure illustrates some typical DNA profiles from a kinship testing case using the SGM*plus*® system. SGM*plus*® analyses 10 loci (20 alleles) and includes a sex marker (amelogenin). This system has a discriminating power of less than 1 in 1 billion; in other words, the chances of failing to distinguish between two unrelated individuals chosen at random is less than 1 in 1 billion. Each locus consists of two alleles, and when they are different (heterozygous), which is often the case, two peaks will be seen in the EPG. Where the same allele is inherited from each parent (homozygous), only one peak will be detected which will be taller due to the increased signal from the two identical alleles.

In Figure 9, four DNA profiles can be seen – from top to bottom, possible father 1, possible father 2, child (a son), and mother. At the far left of each profile, the dotted peaks indicate the sex of the donor – two peaks (from the X and Y chromosomes) for the three males and a single homozygous peak (XX) from the mother. Unless there has been mutation (which is very rare but is factored into any comparison), all of the alleles in the child's profile must have come

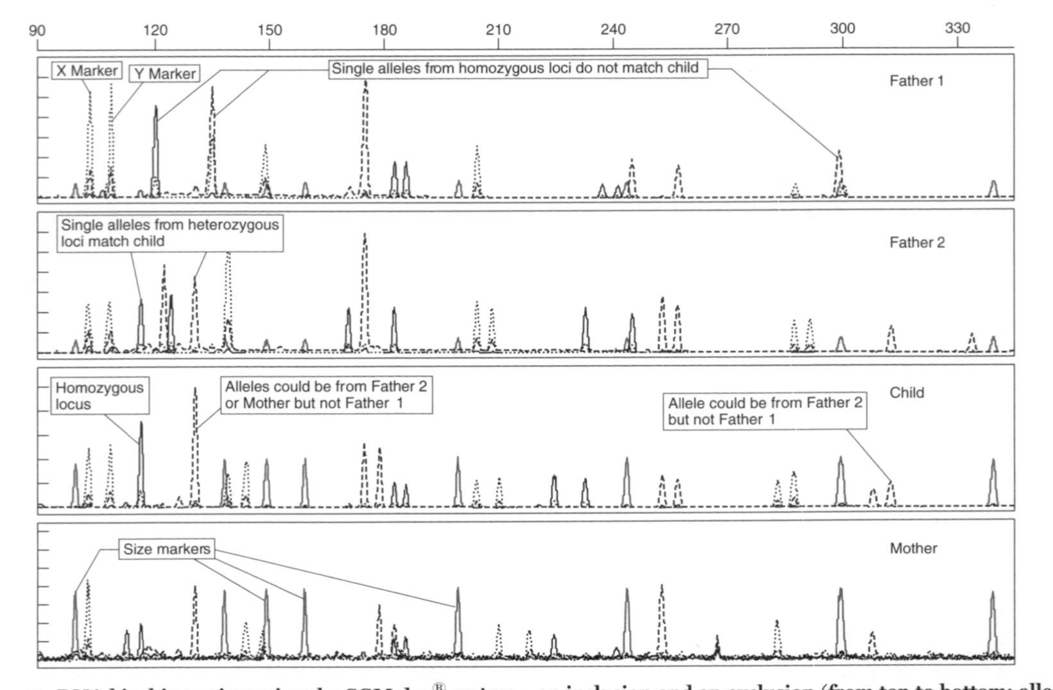

9. DNA kinship testing using the SGMplus® system – an inclusion and an exclusion (from top to bottom: alleged father one, alleged father two, child, mother)

from either the mother or father. As we move from left to right, the first black peak in the child's EPG shows he is homozygous for this locus. The mother shows two peaks at this position and is therefore heterozygous. One of these alleles matches the child, as would be expected. If we consider the EPGs from the putative fathers, father 2 is heterozygous and one of these alleles matches the child; father 1 is homozygous but the allele is different from the child's (the single peak is in a slightly different position). This indicates that father 2 could be the true biological father and provides the first indication that father 1 may be or may not be. The next peaks in the EPG are dashes, representing a second locus. The first large dashed peak at the left of the child's EPG indicates he is homozygous at this locus (D3S1358), as is the mother. Father 2 is heterozygous and one of the alleles matches, therefore he must be included at this stage. Father 1 is homozygous but has a different allele, and this further supports the contention that he is not the true father. This process is continued until every locus has been compared, and at each stage evidence accumulates that father 1 is different (at six loci) and that father 2 could be the true father. Following the analysis, a calculation of the probability of father 2 being the biological father of the child compared to any male in the local population can be reported.

The second example is typical of a DNA profile obtained from a stain in a sexual offence. In most cases, such stains come from mixtures of body fluids from the parties involved, typically semen, vaginal secretions, and/or saliva. If this is not established from the initial analysis of the stain, it may be inferred from the fact that there are more than two true alleles present at any one locus, or from the location where the stain was found, such as on the skin or clothing of the victim. In this case, the DNA profile derives from staining on a vaginal swab taken from the victim of an alleged rape. The top EPG is from a differential extraction of semen from the swab, the middle one is from the suspect, and the bottom EPG from the victim. DNA from the cells of the victim will be found on the swab, and the purpose of differential extraction is to remove the DNA from any sperm while leaving behind that from the vaginal cells. This can be achieved because the sperm is more resistant to the enzyme used to break down the cells and can then be separated by centrifugation. This process does not always completely work and sometimes a mixed profile is obtained. In order to interpret this profile some assumptions have to be made. The main one is that any alleles found which match the donor are from the donor and that only alleles which differ from the donor can be reliably attributed to another individual. It is clear that the DNA in the stain is from a male, since both the X and Y markers have been detected. By comparing the EPG from the stain with that of the victim, you will see that there are a number of peaks which could not have originated from the victim, some of which are indicated. If these peaks are then compared with the profile from the alleged offender, you will see that these match this individual. There are other peaks present which are shared by each of the individuals involved. In summary, DNA has been found on the vaginal swab from the victim which could not be from the victim and matches the alleged offender. The question then becomes: what is the significance of this evidence?

The evaluation of the significance of a matching profile can be done in a number of different ways, but in all cases relies on the frequency of the profile obtained. This requires understanding of population genetics and statistics that allow estimation of allele frequencies and hence genotypes. This takes into account issues such as population size, chance of co-ancestry, and sample size. Since the alleles that lead to the genotypes are independently inherited, the probability of the profile is the product of the individual genotypes. An example from a straightforward unmixed profile is illustrated in Table 9. The chance of two unrelated people sharing this profile, the match probability, is the reciprocal of the profile frequency and in this case is 1 in 7.09×10^{-13}.

It will be clear from these calculations that a matching DNA profile can be used to attribute body fluids or tissues to an individual with great confidence.

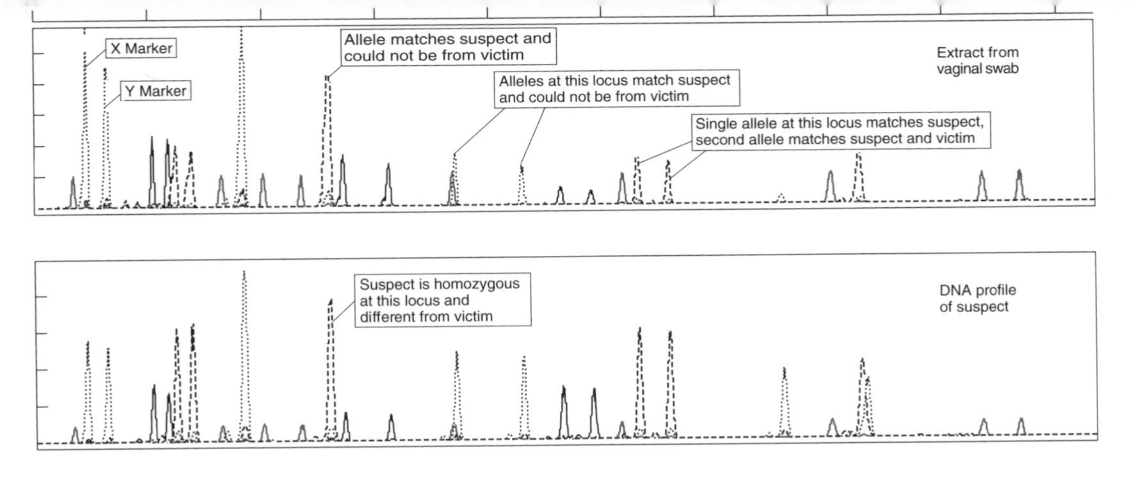
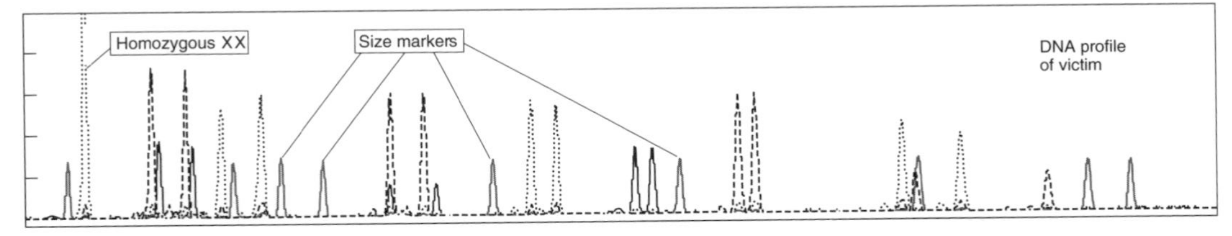

10. DNA mixed stain interpretation (top to bottom: sperm fraction, male suspect, female victim)

Table 9. DNA genotype frequencies. For the purpose of illustration, the frequencies for father 2 in the example above have been used

Locus	Genotype	Genotype frequency
D19	14, 14	0.129
D3	17, 17	0.032
D8	14, 14	0.256
VWA	17, 17	0.073
THO1	9, 9.3	0.077
D21	30, 30	0.053
FGA	21, 22	0.058
D16	11, 14	0.081
D18	15, 18	0.024
D2	20, 20	0.020
Complete profile	All of the above	7.09×10^{-13}

Evaluating evidence

One of the reasons why DNA provides such powerful numerical evidence is that it combines two of the most important elements of science. First, it uses empirical data about the world around us, in this case about population genetics and relationships between people. Second, these data are interpreted using objective statistical methods. As individuals, we are constantly making judgements about the world around us, but it is well known that our subjective judgements and assessments of probabilities are flawed in many instances. Furthermore, our judgements are subject to certain innate biases (see Chapter 6 for a further discussion on this issue). An understanding of statistical probability highlights some of these flaws. A good example of our poor ability to estimate probabilities is the 'gambler's dilemma'. Following a long run of red in a roulette game, what should we bet on the next spin of the wheel? Many people believe subjectively that black is more likely, but since each spin of the wheel is independent of the previous one, the probability for black or red remain the same. A similar issue can occur with the tossing of a coin. We tend to see what we consider to be a pattern from a very short run of tosses that are statistically insignificant – say four heads in a row. Again, the odds of obtaining a head or a tail remain the same. Such dangers also apply to the interpretation of forensic evidence, and therefore it is much better to rely on statistical probabilities than our subjective judgements. Paul Kirk considered probability to be the 'keynote of the interpretation of all physical evidence'.

There are a number of different ways of estimating the significance of DNA profiles, such as the profile frequency (how often it is likely to be found in a given population) or the match probability (the reciprocal of the frequency), which both give a guide. However, there are difficulties with frequencies and match probabilities also have their limitations. One of these is that when match probabilities become very small they can be easily misunderstood. How does one interpret a match probability of 1 in a billion with reference to the UK population, which is around 60 million? A second difficulty is that a very small match probability in relation to a person who is innocent can be misinterpreted as implying that the person is guilty if the evidence is found on them. This misinterpretation is known as the 'prosecutor's fallacy', or transposed conditional. Another statistical means of evaluating evidence used for DNA profiles is based on a theory invented by the Reverend Thomas Bayes, in 1764, which uses a ratio of probabilities called the likelihood ratio. In the context of forensic science, the important probabilities are the probability of the evidence if the prosecution's proposition is true (i.e. the person is guilty) and the probability of the evidence if the defence proposition is true (i.e. the person is innocent).

Bayesian evaluation of evidence avoids some of the difficulties which can be encountered when the relative frequency of the evidence is used. This can be illustrated using a well-known theoretical example. Suppose a rape has been committed in a town where there are 10,000 men and for other reasons we can be confident one of them committed the crime. From the crime scene, traces of minerals are found which connect the offender to a local mine where 200 men work. When a suspect is arrested, similar traces are found on his clothing. What is the significance of this evidence? To illustrate this case, we need to assume that all of the men in the mine will have similar mineral traces on them. Although this is not necessarily the case, let us say it is valid for the purpose of this illustration. There are 9,999 men in the town who are innocent and 199 of them work in the mine. We can estimate the probability of finding this evidence on an innocent man as 199/ 9,999; approximately 0.02. This implies that the evidence is uncommon and therefore significant. But what is the probability of finding the evidence given that the man is innocent? Since we can expect all of the men who work in the mine to have the evidence on them, we can estimate this using the ratio 199/200; approximately 0.995. This gives a very different impression since such evidence is extremely common on the mineworkers. These probabilities illustrate the prosecutor's fallacy; in fact, the probability of the mineral traces given that the man works in the mine indicate he is much more likely to be innocent than guilty. The use of Bayesian reasoning for evaluation of evidence is not universally accepted and has presented some difficulties in court due to its complexity. However, most statisticians and many forensic scientists consider it to be the most effective method.

DNA databases

The world's first DNA database was created in England and Wales in 1995 and since then DNA databases have been implemented in most developed countries in the world. The rationale for most databases follows the same general principles: most crimes are committed by a minority of individuals; many of these individuals reoffend; most criminals are involved in a wide range of crimes; many individuals involved in serious crime are also involved in minor crime. Therefore retaining DNA (and fingerprints) from convicted persons means that offenders who already have convictions can probably be identified and arrested more quickly.

The nature of DNA databases in different countries varies considerably due to differences in legislation, police procedures, and the particular type of DNA profiling involved. In some countries, DNA is retained only for certain serious offences. In other countries, samples are stored but are eventually destroyed. In England and Wales, until recently, all DNA samples were retained indefinitely irrespective of whether the person was convicted or not. However, in December 2008, the European Court of Human Rights ruled against this, requiring a change in the law. Indefinite retention of DNA from non-convicted persons has never been legal in Scotland. The UK has the most permissive legislation in the world and is the only country that allows retention of DNA from non-convicted persons. It is also the only country that allows speculative searching of the national DNA database. For example, if an individual is arrested (irrespective of whether they are convicted) for one alleged crime, the database can be searched to see if their DNA profile is present in connection with other incidents. In other countries this is not the case. For example, in the USA, the police must have reasonable cause for suspecting that the individual has committed a particular crime. Furthermore, the range of crimes for which DNA can be taken from individuals in the USA, and other countries such as Australia, varies from state to state. The extent to which DNA databases can be used appears to be linked to whether ownership of the data is with the police or the judiciary. Where the data are owned by the police, legislation tends to be more permissive, as in the UK. In continental Europe, where the data are largely owned by the judiciary, use of DNA data is more restricted. Scotland is a good example of a country that balances ownership of the data by the police with judicial control of certain aspects of its use.

It is very clear that DNA databases contribute positively to the investigation and prosecution of crime, but very little research has been carried out to demonstrate how they contribute. Such research could answer some of the following questions: what role does DNA play in the detection of crime?; what role does DNA play in the prosecution of crime?; is DNA more or less effective for some crimes compared to others?; is crime detected more quickly when DNA is involved? At the moment, we only know the answers to these questions from anecdotal evidence.

DNA profiling provides a powerful tool for the investigation of crime because blood and body fluids are transferred in a wide range of crimes and it can analyse extremely small quantities. The method of analysis uses multiple tests that can be carried out quickly and economically on a single stain. The analysis is highly discriminating and provides very strong evidence. This means that offenders can be rapidly identified if their profiles are stored on a database, and intelligence-led screens (mass screens) can be used to eliminate large numbers of individuals, many of whom may be volunteers. Furthermore, an offender who is not on the database but who has a relative on it may be identified indirectly, since they will have a similar DNA profile (familial searching).

Chapter 6
Prints and marks: more ways to identify people and things

Marks (or impressions) are caused by a pattern from one item being transferred to another. This could be a shoe mark, a finger mark, or, less obviously, the pattern of striations on a plastic bag made by a tool in the manufacturing process. Firing pins in guns (see Figure 11), as well as saws, tyres, screwdrivers, and feet can all leave marks that can be used to identify the general type of object that made them (a shoe, a tyre) and sometimes even the specific object. This chapter uses fingerprints and shoe marks to illustrate the general characteristics of marks evidence, the principles involved in their examination, and how the evidence is evaluated.

Fingerprints and shoe marks are the most important and frequently encountered evidence of this type. In the view of most marks examiners, identification of an object (or an individual by fingerprints) can be done unequivocally, that is, with 100% certainty. We have everyday experience of marks and have ourselves on occasions made such judgements: who muddied the kitchen floor – a small boy (a shoe mark) or a small dog (a paw mark)? Understanding the application of marks to the investigation of crime is an extension of this everyday experience. Less obvious is how such marks can be used in intelligence databases to analyse crime patterns, or determine if someone should be charged with possession of drugs or possession with intent to supply drugs.

Marks can be visible (patent) or invisible (latent) and require specialist optical, physical, or chemical techniques to visualize them. They can be made in a variety of substances: mud, blood, dust, sweat, soot – referred to as 'negative' marks; or by transferring a material to another surface – 'positive' marks. A shoe stepping into a pool of blood can leave a negative mark in the blood followed by positive marks on the floor walked upon.

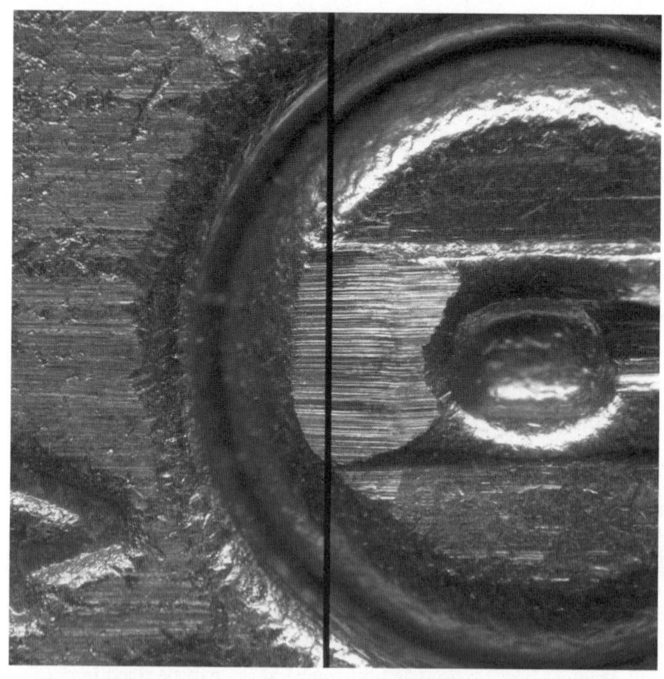

11. Striations on bullet casing. This image illustrates a recovered cartridge and test-fired cartridge under a comparison microscope showing matching striations on both cartridges indicating the bullets have been fired by the same weapon

A characteristic feature of many marks is that one can identify the type of item that made the mark from examination of the mark's general characteristics. The appearance of the mark generally represents the shape of the item, its main features, and their spatial arrangement. Shoe marks, finger marks, and tyre marks are readily identifiable often from a cursory visual examination, though this is not always the case. In addition, marks can provide further information about the type of item that made them. Not just a shoe but a particular type of shoe (a Reebok Classic, say), or a saw with a certain number of teeth per inch, or a screwdriver of a certain blade width. This provides the opportunity to derive inceptive information or intelligence for the investigator in the absence of the item or a suspect. Knowing what the item is enables the police to look out for it. If information from large numbers of marks is compiled as a database, it can be used as intelligence to link crime scenes. Such databases are widely used in fingerprints and we will discuss their application for shoes later.

Despite the wide range of objects that make marks, the processes and aim of any comparison are very similar: to determine whether there is a connection or otherwise between the mark and the reference object, and if so to what degree. We have already discussed the nature of identity and how DNA matches are evaluated in probabilistic terms, but this process is not used for marks identifications. Instead, following the gradual accumulation of similarities between the mark and reference item, there comes a point when the examiner decides that it matches this reference item (shoe, tool, finger) and no other. In scientific terms, this is to say the least an odd conclusion since it cannot possibly be justified on logical grounds. For such a, conclusion to be logical, the examiner would have to have compared the mark with all other reference marks currently in existence. Yet most examiners are prepared to state such a conclusion, and the courts are generally content to receive such conclusions since they much prefer evidence that is clear and easy to understand. We will consider this issue further under fingerprints. The quality of the mark, that is, the amount of information and detail present, is crucial to the determination of a match and the degree of matching. The comparison process routinely carried out in forensic science laboratories follows a similar general procedure for most marks examinations and addresses the following questions:

- What are the characteristics and features of the mark?
- Can they be observed in the reference item?
- Do the characteristics and features of the mark match the reference item?
- Do any characteristics appear to be different?
- How significant are the matching features?
- Are any differences significant (should the mark be eliminated or are the differences explicable)?
- How significant is the match or elimination?

Another important consideration is how to evaluate differences between the mark and reference item. Such differences do not necessarily mean the item should be eliminated. Tools, shoes, and tyres wear as they are used and will continue to wear after they have deposited any marks. This can result in new features, for example as a result of damage, which are not in the mark or loss of pattern features as the mark wears. Such differences must be evaluated by the examiner on the basis of their knowledge and experience. For the most part, this is not an issue with fingerprints, as we will see.

Fingerprints

'Fingerprint' is a byword for identification and identity. It is widely used in everyday language to refer to any set of characteristics that define an object, activity, or even style. Of the differing types of impressions encountered in forensic work, fingerprints are a particular case since they are biological in origin, not manufactured, and because they directly identify individuals, as does DNA. Fingerprints have been used to identify victims and criminals in the UK for over 100 years, and there are records from around 8 million individuals stored on the national database (IDENT1). Fingerprints arise from a particular type of skin called friction ridge skin found on the fingers and palms of the hands, and the soles of the feet. This skin has patterns of ridges and furrows which can be transferred in sweat to other items when they are touched, for example when picking up a drinking glass. The origin of these patterns is genetic and they develop in the foetus in the womb. However, they are also subject to non-genetic influence since identical twins, who have an identical genome and DNA, have different fingerprints.

The examination of fingerprints is based on the detailed characteristics which form the patterns of ridges and which can be compared. Unless damaged in some way, these skin patterns persist throughout life, providing a powerful biometric which can be used to identify individuals and store records in databases. Fingerprints are routinely used in most countries around the world to check that an individual who is arrested is who they say they are and whether they have come to police attention previously. They are also used as a means of identity in non-criminal situations such as general security in buildings and access to computers. However, fingerprints are perhaps best known as a means not just of identifying individuals but also of linking them via marks to a crime scene or an item, such as a weapon or vehicle, that may inculpate them. Despite this widely known fact and that transfer of fingerprints can be easily prevented by wearing gloves, they are still found at crime scenes around the world in vast numbers, and the examination of fingerprints is one of the most important and valuable areas of forensic science.

However, in many countries, fingerprint examination is separate from the rest of forensic science for reasons that are perhaps historic (and rather too lengthy to go into here) but are no longer justifiable. The current physical, methodological, cognitive, and cultural separation of fingerprints from other areas of forensic science, in my view, serves neither the interests of justice nor the long-term interest of fingerprint experts. Recent developments in Scotland, where all specialist forensic services reside in a single organization, are likely to serve criminal justice more effectively.

History

The idea of fingerprints as an identifier has been around for millennia; for example, marks were used historically to identify ownership of clay pots in the ancient world.The modern history of fingerprint examination began around 100 years ago, and the first case in the UK to use marks from crime scenes was in 1902.

In 1893, the UK adopted a system to identify repeat offenders which used a combination of Alphonse Bertillion's anthropometric system and Francis Galton's fingerprint system. In 1900, the Bertillion system was abandoned as impractical and ineffective, and fingerprints were adopted as the sole means of identification. Henry Faulds, a former student at Anderson's College in Glasgow (now the University of Strathclyde), was the first person to propose the use of fingerprints in the investigation of crime, in his paper published in the scientific journal *Nature* in 1880. Numerous others, particularly Francis Galton, the English polymath and cousin of Charles Darwin, were responsible for the development of procedures for examining fingerprints and encouraging their use. Although there have been some alterations over the years, the processes developed in the early part of the 20th century essentially remained in use until the development of computerized automatic fingerprint identification systems (AFIS) in the 1980s and live scan technology (digital capture of fingerprints) in the 1990s.

Features of fingerprints

Until now we have used the terms 'print' and 'mark' interchangeably, but there is an important convention in the use of this terminology. Both terms refer to patterns deposited onto surfaces, but the word 'print' refers to impressions made by known sources. When impressions are taken from an individual as records and to identify the individual, these are referred to as 'fingerprints'. The sets of all ten finger impressions taken for police records are referred to as 'ten-prints' in the UK. Impressions from fingers left on objects are referred to as 'finger marks', or simply 'marks'. This is an important distinction between circumstances where the source of the impression is known and those where it is unknown but can be inferred following examination.

Marks are generally formed from sweat left on the surface of the ridges. The deposited residue is composed mainly of water but also contains proteins, amino acids, fatty acids, inorganic salts, cholesterol, and squalene. The amount of residue left by an individual depends on a large number of variables, such as the condition of the skin and the diet, age, sex, and physical condition of the donor. The surface on which a mark is deposited is also important. Marks left on porous surfaces such as paper may remain for decades because the residue is absorbed and fixed. In contrast, marks on a non-porous surface are prone to abrasion and their persistence is significantly influenced by the subsequent handling of the item.

Three main patterns of ridges are discernible: loops, whorls, and arches (see Figure 12), and these can be used to classify prints and marks or rapidly exclude a mark that has a distinctly different pattern.

The ridges are not continuous but composed of various arrangements and characteristics known as *minutiae* including bifurcations, ridge endings, and other related phenomena. The ability to characterize a fingerprint and identify a donor is based primarily on the sequence of minutiae in the print and mark. These sequences are extraordinarily discriminating (like DNA) and the amount of information in a mark is often highly redundant. The same mark can be identified in a number of different ways and it is

12. Fingerprint ridge patterns. This shows the three main categories used: loop (right), whorl (left) and arch (top). Non matching patterns can be used to rapidly eliminate a mark

not uncommon for different fingerprint experts who have reached the same conclusion about a mark to do so using a different sequence of minutiae. For an identification, there must be a significant number of minutiae *in sequence and agreement* and no significant or inexplicable differences. Two of the main minutiae are illustrated in Figure 13, together with a comparison of an unknown mark with a reference print.

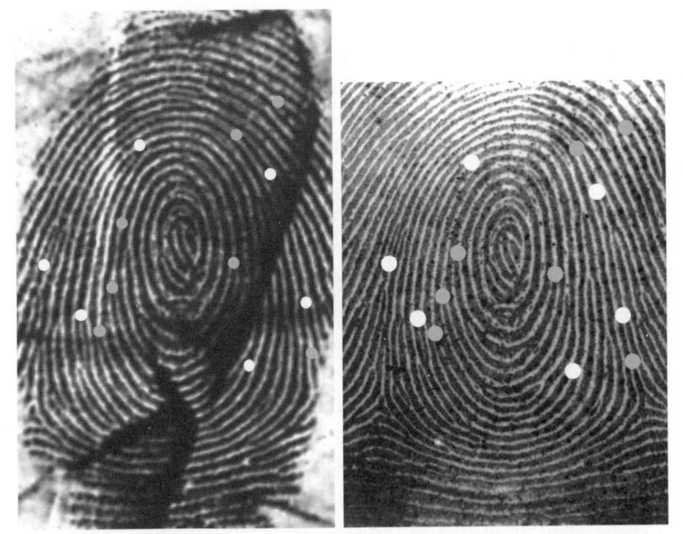

13a. Fingerprint minutiae: bifurcation (top) and ridge ending (below)
13b. Fingerprint comparison of mark (left) and print (right) showing bifurcation and ridge endings in sequence and agreement

In addition to the pattern of the mark and sequence of minutiae, other features such as the pattern of minute pores on the tops of the ridges can also be used to identify marks.

Recovery of marks

The oldest and most common method for recovering latent marks is the use of fine powders which are applied to the mark by a brush. A number of commercial powders are available in differing materials, particle size, shape, and colour which can be used on surfaces of different textures and colours. Powders such as aluminium, the most common one used, have a tendency to adhere to the fatty components of the residue. These work best on non-porous surfaces and less well on others, but they have the advantage of simplicity and ease of application. There is a wide range of other techniques that can be used for the detection and visualization of latent marks which depend on the nature of the surface on which the mark is deposited and components in the residue of the mark. Lasers and ultraviolet light can be used to promote fluorescence of marks which can then be photographed. Marks can also be enhanced by a range of chemicals that react with specific components in the residue. One of the best-known methods uses ninhydrin, a general protein stain, to react with amino acids in the mark. This is very effective for porous surfaces and is commonly used to develop marks on paper. A common chemical method for non-porous surfaces uses vaporized cynaoacrylate (a constituent of superglue) which reacts with the residue, rendering it visible as pale grey or white marks. These methods are frequently applied in a logical sequence following a well-known publication, the *Manual of Fingerprint Development* published by the Home Office Scientific Development Branch in the UK.

Fingerprint comparison

When comparing marks to prints, there is now widespread acceptance amongst fingerprint examiners of a standard methodology that is denoted by the acronym ACE-V: analysis, comparison, evaluation, and verification.

- Analysis. In this stage, a detailed assessment of the quality and level of information in a mark is made. Issues such as the possible substrate and its influence on the mark are clarified. The effects of distortion and pressure are also taken into account, which enables the examiner to allow for minute variations (tolerances) between the mark and print.
- Comparison. This is the 'side-by-side' examination to establish correspondences (or lack of them) between the mark and print. According to Christophe Champod, a noted academic and experienced forensic expert who advised the Public Inquiry into the Shirley McKie case, the mark should be examined first and the print next, and if the reverse process is used it should be done with great care and specifically recorded in the case notes.
- Evaluation. In light of any correspondences between the print and mark, the examiner makes a judgement: can the mark be eliminated or does it match? If it matches, is the degree of correspondence sufficient to identify an individual? This judgement is an inference and therefore subjective.
- Verification. A second experienced and qualified fingerprint expert independently reviews the comparison and conclusion following the ACE-V protocol.

The outcome of this process is one of three possible conclusions. The first of these is an *identification* (or 'individualization', as it is called in some parts of the world), which means that the mark is attributed to one person to the exclusion of all others. The second possibility is *exclusion*: the person whose prints have been used for comparison could not have made the mark. The third category is *inconclusive* (or insufficient): that is, no judgement can be made about the mark (usually due to poor quality or lack of detail present) for the purposes of a criminal investigation or prosecution. A little reflection on this third category will immediately reveal it as problematic.

Let's consider these outcomes in a little more detail. If there are significant differences between a mark and a print, an individual can be excluded. This is analogous to practices elsewhere in forensic science – exclusion is often a very straightforward business. One critical difference is enough, irrespective of what is being compared (hairs, fibres, paint fragments). Identification takes place when the examiner considers there has been sufficient accumulation of elements (minutiae, etc.) that lead them to form the view that a mark can be attributed unequivocally to an individual. Marks that fall between these two categories are deemed inconclusive, but this must necessarily include a range of marks with different levels of information present. Some of these will contain little or no information: perhaps a pattern and one or two minutiae. Others will contain considerable detail, although not enough for identification, but does this mean they are of no value? In all other areas of forensic science (including marks examination), where evidence falls short of identification but reveals matching features the examiner will comment on the significance of the match. This practice of categorizing non-identifiable evidence as inconclusive is unique to fingerprints, and we will return to this point towards the end of this chapter.

Identification standards

Fingerprint identification standards have been developing over the past 50 years or so, but the trajectory of this development has not always been logical nor based on fact. The situation has also been different in different countries around the world, but there are some commonalities. For a lengthy period, it was generally agreed in the fingerprint world that a specified number of minutiae in sequence and agreement (and no differences) were required in order to identify a mark. This number varied in different countries and even sometimes within one country. Until 2001 in England (2006 in Scotland), the number was 16 minutiae – colloquially and inaccurately referred to as 'the 16-point standard'. This was not in fact a standard but a numerical threshold or bar of convenience for fingerprint experts. This meant that a mark having 16 points could be identified but one that had fewer (other than certain specific circumstances) could not. There was no halfway house – if only 14 points were present, the mark was not identified. Gradually numeric standards were eroded by shifting opinion in the fingerprint world and with the acceptance that there was no logical basis for a numeric standard. A study of considerable influence on this issue was carried out by Ian Evett and Ray Williams in which they identified such variation in the attribution of numbers of minutiae between different experts that obtaining 16 points had very little influence on 'standards'. Amongst other important findings, they established that the 16-point standard was based on false evidence and that there was worrying practice by some experts of using the print as a guide to examining the mark. It is now widely accepted that a non-numeric standard based on a systematic and detailed examination by a trained examiner, taking into account all available features in the mark, is a more logical approach, and this is the situation that exists in the UK, North America, Australia, and the Nordic countries.

The national fingerprint database in the UK (IDENT1) stores ten-prints from almost 8 million individuals and typically enables fingerprint examiners to make 90,000 identifications per year from marks at scenes of crime. Every person in the UK arrested for a recordable offence has their fingerprints taken and stored on this database, which is also linked to the criminal record systems in the UK. The database is a powerful tool for identifying individuals, linking individuals to crime scenes, and linking crime scenes that involve the same offender.

Shoe marks

Footwear marks are used in a wide variety of investigations including volume and serious crime. As well as being useful to link marks and shoes, shoe marks offer the possibility of pre-emptive information (intelligence) and can contribute to an investigation prior to the arrest of an offender by identifying the types of shoes worn or by linking crime scenes. Footwear marks can indicate the position of individuals in a scene and their movements, which may be important to the investigation or prosecution of a case. For example, a shoe mark found near a window may be a good indication that this was the point of entry (or exit) in a burglary. Shoe marks can be recovered in very similar ways to fingerprints. The most common method used is photography, which must include a scale so that the mark can be reproduced for life-size comparison. Latent marks can also be visualized using the same optical, chemical, and physical enhancement processes as were mentioned above. There are a number of important features used in examination of shoe marks: the tread pattern, that is, the arrangement of the various individual elements that make up the sole (lines, circles, squares, logos, etc.); the tread pattern dimensions – the sizes of individual elements and overall size of the pattern; manufacturing features such as mould marks, bubbles, and knife marks from trimming; progressive wear, which results in gradual loss of detail in the pattern as the shoe is worn, and specific damage – individual and characteristic elements of damage such as cuts caused by wear or defects from the manufacturing process.

All of these can be used individually or in combination to examine marks, and this usually involves producing a test impression of the tread. Figure 14 illustrates the comparison of an unknown mark and known shoe print. The larger the area of the tread

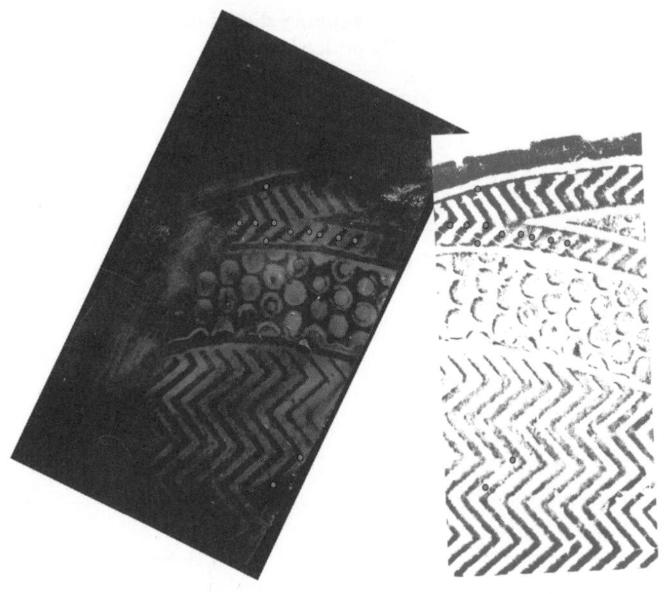

14. Shoe mark comparison. This illustrates the matching features in an unknown mark and a reference shoe print

represented and better the definition, the more information is available to determine an association or exclusion. Non-matching features must be taken into account in this process. Differences that do not have a reasonable explanation should result in the examiner excluding the footwear from having made the mark. A difference in pattern would exclude the shoe immediately and could take place very quickly if the patterns are noticeably different.

In addition to the tread pattern, there are other aspects of the way a mark was made that affect its appearance. For example, the amount of pressure on the tread and the degree of movement involved in deposition can result in distortion of the pattern

Table 10. Typical pattern frequency distribution for footwear marks from a UK police force

Make	Model	% Marks/Footwear
Nike	Max Ltd	10
Nike	Max 95	8
Adidas	Campus	3
Reebok	Classic 1	3
Nike	Max 90	3
Nike	Court Tradition	2
Nike	Tuned 8	1.5
Lacoste	Camden	1.5
Various	Ripple	1.5
Adidas	Country	1.5
Nike	Max Classic	1.5
Other patterns		Less than 1%

(as in fingerprints). Unstable surfaces such as soil or sand can also lower the mark quality.

Footwear intelligence databases

Intelligence databases containing information on footwear have been in use for many years. These databases have two principal aims: to link crime scenes and make associations between footwear from offenders and marks from crime scenes. The basis of the database is that specific tread patterns are associated with particular makes and models of shoe. The database does not tell you if the shoe made the mark but which type of shoe the mark may have come from: for example, if the mark belongs to a particular brand of trainers. This information may lead the police to a particular suspect who owns such shoes or inform searches of suspects' homes. Following seizure of the shoe, a full comparison is made by a marks examiner to determine if this shoe could have made the mark. All intelligence must be considered in context – the same shoe type at two different scenes may not indicate a link if the shoes are very common. Such databases can also help determine how common certain types of footwear are, which helps to assess the significance of intelligence matches and assessment of the evidential value of a match. Table 10 shows typical frequencies for a number of footwear types in the UK. Even the most common is only encountered in 10% of the cases recorded.

Evaluating marks

In Chapter 5, we considered various methods for evaluating the significance of DNA evidence, all of which used a statistical approach. This follows the scientific principle that all such judgements contain a degree of uncertainty no matter how small, and the use of statistics allows this to be taken into account. This is known as probabilistic evaluation. It will be clear from the above that this approach is not used for marks or fingerprint evidence where examiners make judgements about identity that are categoric, that is, certain. This approach has attracted considerable criticism from outside the forensic community and from significant numbers inside the community.

This debate has three important components: validity (is using categoric judgements for any evidence types justifiable?); feasibility (can probabilistic methodologies be developed for evidence types which traditionally use categoric methods?); and necessity (is there a need to use probabilistic methodologies even if they were available?).

It is increasingly obvious that marks and fingerprints will soon be amenable to probabilistic modelling and decision making. The issue of feasibility is, for the most part, a red herring. Current research in relation to fingerprints is well advanced, and I would anticipate probabilistic evaluation being used in courts in the UK in the next few years. Given this, we are left with the issues of validity and necessity, of which there are perhaps three schools of thought: universalists, pragmatists, and traditionalists. Universalists take the view that all evidence should be evaluated by probabilistic means and that this is the only rational way to approach the issue. Traditionalists consider categoric evaluation valid, are resistant to probabilistic approaches, and are frequently unaware of the methodologies that may be used for probabilistic evaluation. Pragmatists take the view that few professions can claim to have universal standards and that there is such inherent variation in criminal justice systems that, although probabilistic evaluation is desirable, it is not necessary in every case.

Confirmation bias

Confirmation bias is another topical debate in forensic science and will continue to be so until there is some resolution of the issues raised. It has particularly come to notice in relation to fingerprints but has direct relevance to other areas of forensic science, especially marks. It is a widely known fact of human behaviour that an individual's expectations can influence their perceptions and judgement. When you look at the night sky for a constellation, you are looking for a pattern that you expect to find; there is no pattern there – you are imposing the pattern on a random distribution of stars in the sky. Not only will you see the pattern if you expect it, but you will have trouble seeing things that do not fit that pattern and may not perceive these. This is why it is so important to examine a mark before a print.

In forensic science, cases are not examined in a vacuum, a certain amount of information is necessary for the examiner to have a general understanding of the case – when and where did it take place, who are the victims and suspects, what other information is already known? It is perhaps this last piece of information that is most problematic. The purpose of the information is to enable the examiner to carry out an efficient and effective examination; one that takes into account the circumstances of the case and enables them to determine examination needs and priorities. There are two problems with this. First, some of this information can set up expectations, and this needs to be carefully handled. Second, together with the relevant information comes a great deal of irrelevant information which is potentially prejudicial. As a theoretical example, a shoe is submitted for examination in a homicide case in which shoe marks in blood have been found at the scene. This will need to be examined by two scientists, one to deal with the blood and DNA profiling and one to carry out the marks comparison. The fact that the DNA scientist finds that the DNA profile on the shoe of the suspect matches the victim is completely irrelevant to the examination of the mark in blood. For the marks scientist to know this in advance potentially sets up expectations of the outcome of her examination, that is, that it should match because the DNA matched. There is no doubt about these effects; they have been widely reported in the literature for many years in domains other than forensic science. The issue that is often misunderstood, especially by fingerprint experts, is not whether such effects exist or not (they do) but whether they are 'under control' by using appropriate methods. It is one of the triumphs of the scientific method that it has allowed us to know that these effects exist, and it can help prevent these prejudices and preconceptions. The latter can be achieved by using a suitable scientific methodology, careful recording of information, and detailed recording of the rationale for decisions.

But there is considerable disagreement about the significance and influence of such phenomena. Are such effects possible in fingerprints – is the background to the case likely to influence the outcome of a comparison? Some would say yes, and cite a landmark study by Itiel Dror in 2006. By using the same sets of matching finger marks and prints and having them examined (unknowingly) by the same experts at different times, Dror was able to demonstrate that most of the experts made inconsistent decisions. Only one of the five experts involved made the same decision as previously. Three of the four experts excluded the previously matching mark and one decided the examination of the mark was inconclusive. These are devastating findings that fundamentally question the standards of fingerprint experts and their procedures, but they are generally in line with what a psychologist would expect of human behaviour. The question is: what is the significance of this study in terms of forensic practice? Two further studies have been carried out, one on fingerprints and one on shoe marks, neither of which found observer or context bias as they are known.

We have considered marks and impression evidence, with particular focus on shoe marks and fingerprints, and explained why they play such a crucial role in the investigation of crime. In general terms, marks and impression evidence are at something of a watershed due to criticism of some of the methodologies used and the dangers these may present. It is likely that we will see significant changes in procedures in this area which will require large-scale alterations to working practices and the training of those involved. This is necessary for them to keep pace with developments in other areas of forensic science and maintain their essential contribution to the investigation and prosecution of crime.

Chapter 7
Trace evidence

In many ways, the concept of trace evidence epitomizes forensic science. The idea that tiny fragments of materials, invisible to the naked eye and therefore unknown to those involved, can be used to investigate crime is a powerful one which catches the imagination as well as being of practical value. Fibres, hairs, paint, glass, and explosives traces are examined by forensic scientists in a wide range of crimes from burglary to terrorism. The distinctive characteristics of trace evidence are its microscopic size or minute amount (and therefore its 'invisibility'), its ability to transfer readily from one item to another, and that it is subsequently lost from the item following that transfer. Many of the issues we will explore and the difficulties that arise in the examination, analysis, and interpretation of trace evidence are consequences of these characteristics. The minute amounts involved require specialist techniques to recover and analyse the evidence, as well as stringent precautions to prevent accidental contamination at scenes and in the laboratory. This chapter describes what most forensic scientists would refer to as 'contact trace evidence', and we will consider the principles that underlie trace evidence examination, some of the scientific techniques used, how the value of the evidence is assessed, and the importance of trace evidence in police investigations.

Trace evidence can come from a bewildering range of sources: natural and synthetic. For the most part, we will confine ourselves to fibres, paint, and glass as being the most common types of trace evidence encountered, although we will touch on other types. The defining characteristic of trace evidence is its small size. There is no agreed size at which materials become trace evidence. The larger fragments can often be seen by the naked eye, such as single fibres and some paint and glass fragments, although we could not search for such fragments effectively using the unaided eye. The reasons why these fragments are so small differs depending on the evidence type. For glass, it is a consequence of the material being shattered into fragments during the commission of the crime. For fibres, it may be due to fragmentation of synthetic fibres (which can be very long) or due to the small size of individual natural fibres such as wool or cotton. The next important characteristic of trace evidence is that it is easily and often quickly lost, almost as easily as it is transferred. How long the material stays in place is usually referred to as its 'persistence'. The twin concepts of transfer and persistence are critical to the understanding of trace evidence, as we established in Chapter 1

Fibres

Textiles are all around us, in our homes, cars, and workplaces, in the form of clothing, upholstery, and fabrics of all kinds. Most fabrics are mass produced from natural or synthetic fibres. We know from our everyday experience that fibres readily transfer from one item to another – light fibres show up easily on dark clothing. When a fibre from one item is found on another, two phenomena, transfer and persistence, are involved. In the investigation of crime, fibres can be used to associate individuals with other individuals or crime scenes, vehicles, or items associated with a crime. This would include, for example, connecting an individual to a particular seat in a vehicle, with a balaclava which was used in an armed robbery, or with the clothing of a victim who had been physically attacked. Fibres can also be recovered from weapons such as knives or guns and from vehicles in 'hit and run' cases. We need to make an important distinction between the evidence (the fibres) found and the inferences which may be drawn from them. Strictly speaking, we are not linking individuals or scenes but clothing, items, or fabrics, and the significance of any finding will depend on the detailed circumstances of the case. However, fibre examinations allow us to test many investigative hypotheses and are therefore valuable for this purpose.

Recovery and examination of fibres is a laborious and painstaking process that rarely yields conclusive evidence, and can slow down the examination of other types of evidence which are potentially more fruitful. In a homicide where blood has been shed, any blood found is likely to yield better evidence as it can be attributed with great confidence to its source by DNA profiling. In such a case, fibres may be recovered (which in itself takes a great deal of time) but are unlikely to be examined. In cases where it is accepted that the parties involved have been in contact, or where they have a prior association which could explain the presence of fibres found, a fibres examination would not be carried out. This happens in many cases where the individuals involved are related, share the same home or workplace, or were in contact prior to the incident, for example in a pub or club. Generally, one can only comment on fibres evidence when fibres have been found. In rare instances, it may be possible to draw certain inferences from the absence of fibres, but this is difficult due to the number of imponderable factors involved. Prior to an examination taking place, a great deal of information is required by the scientist to assess the likelihood of finding fibres and their potential significance to an investigation. In doing so, the scientist will pose the following question: how likely is it that fibres have been transferred that are likely to be recovered and could provide evidence of value? Strictly speaking, this question can only be answered by carrying out the work, but in most cases reasonable predictions of the outcome can be made on the basis of knowledge, experience, and the particular details of the case. Table 11 considers the factors that need to be taken into account in this process.

Table 11. Determining the potential value of a fibres examination

Is the integrity of the items sound?	Are there any issues of contamination? Have the items been recovered, packaged, and sealed appropriately?
Does the donor item shed fibres?	Fleecy items shed fibres more readily than smooth items. Natural fibres tend to shed more easily than synthetic fibres.
Will the recipient item retain fibres?	Very smooth surfaces will lose fibres more quickly than rough or fleecy surfaces.
Has there been sufficient contact to transfer fibres?	The longer the contact, the bigger the area, and the higher the pressure involved, the more likely it is that transfer will have taken place.
Was the recipient item recovered in time to minimize loss of any transferred fibres?	Most fibres are lost rapidly after transfer, especially if this is to the clothing of someone who is moving. Fibres on stationary items, e.g. car seats, will remain in place for much longer.
Can the fibres be recovered?	The fibres need to be sufficiently coloured to be seen under a low-power microscope and contrast sufficiently with the donor item. It may be impossible to find blue fibres from one garment on another blue garment even if they are there.
Are the fibres likely to provide significant evidence?	Some fibres are so common in the environment that they are likely to be found on most items and therefore mean very little. Black cotton and cotton from blue jeans are examples of this.

Once the items have been taped, the tapings are searched for a 'target fibre'. In a typical clothing examination, this will be a single fibre type (clothing often contains more than one fibre type) from one item that has good prospects of being found if present and of potential evidential significance. This usually means a fibre of a recognizable colour that can be separated from the background. The tapings are then systematically examined for fibres similar to the target, and any found are removed and mounted individually on microscope slides. This can take many hours and can be like looking for the proverbial needle in a haystack. Each fibre is then compared under a high-power comparison microscope. This microscope (see Figure 15) allows the examination of the recovered fibre and control sample simultaneously under the same lighting conditions.

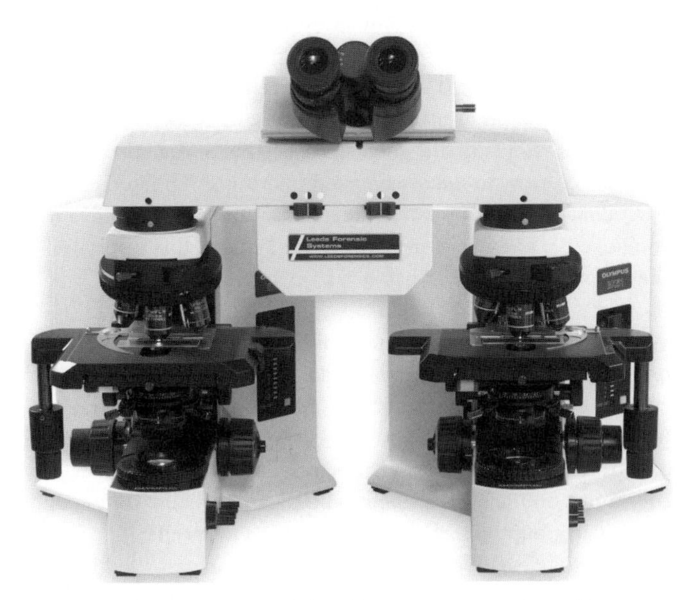

15. Comparison microscope. This allows the examination of the recovered fibre and control sample simultaneously under the same lighting conditions

Many of the fibres recovered will not match and will be discarded as this first stage is a fairly crude process. The features examined during the comparison process include colour, fibre type, and the cross-sectional shape of the fibre. Fibres that match move on to the next stage for detailed analysis, which includes measurement and comparison of the colour and analysis of the polymers present in synthetic fibres. Most natural fibres can be identified on the basis of microscopy alone. The colour of the fibre is analysed and compared by microspectrophotometry. A microspectrophotometer is a specialist microscope with a spectrophotometer attached. This measures the colour spectrum of the fibre (illustrated in Figure 16) and allows more objective comparison than is possible using normal high-power microscopy. The process is very discriminating, but in many laboratories further analysis is carried out on the dyes in individual fibres, which may be only a few millimetres long, by thin-layer chromatography (TLC; see Chapter 8 for more details of this technique). Over 7,000 different dyes are produced worldwide and these are used in many combinations. This test can discriminate fibres that are indistinguishable in the other tests such as comparison microscopy and microspectrophotometry. TLC allows separation and comparison of the individual dye components, and if sufficient detail is obtained, this associates the dye with a particular batch.

Wool, cotton, and many vegetable fibres can be readily identified on the basis of microscopy alone, but synthetic fibres require chemical analysis to confirm their type. The most common method used in forensic science laboratories is Fourier Transform Infrared Spectroscopy (FTIR). Infrared spectroscopy can be used to identify substances on the basis of their molecular vibrations. Different molecular groups absorb infrared radiation at specific wavelengths, resulting in a spectrum that can be used to identify the polymer type present. In FTIR, a mathematical process (Fourier transformation) is used which allows faster collection and analysis of the data. From this analysis, the scientist can identify the type of fibre and in many cases the presence of other polymers which allow subtyping of the fibre.

The above processes have to take place for every suitable item involved and can result in the recovery and examination of many

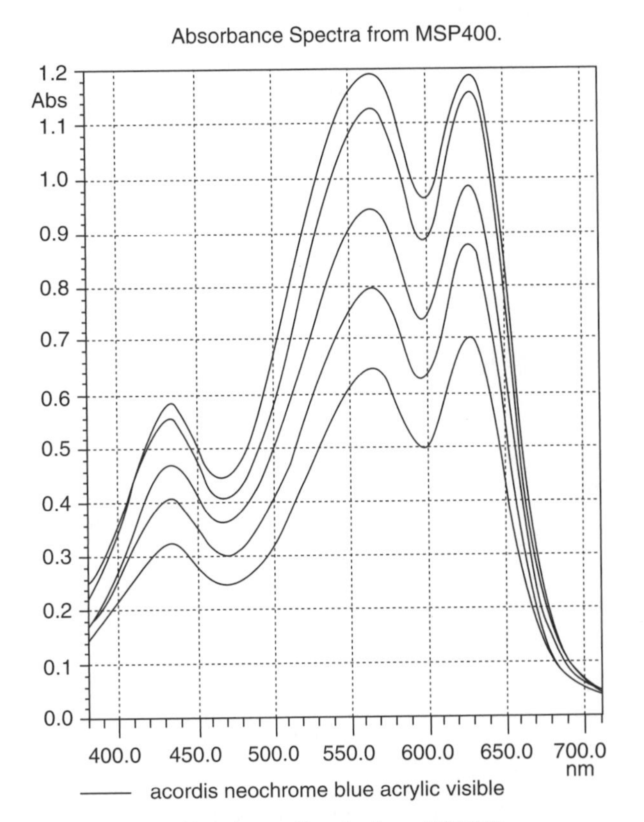

16. Colour spectra of blue acrylic fibres. Examination of such spectra allows the matching or elimination of individual recovered fibres when compared to a control sample from a garment

thousands of fibres and take many weeks. The final stage is the interpretation of the evidence. In addition to the detailed findings and specific context of the case, the main factors considered here are set out in Table 12:

Table 12. Evaluating the significance of fibres evidence. Factors that are considered in determining the significance of recovered matching fibres

Numbers of fibres	The more matching fibres found, the more confident one can be of direct contact. Very small numbers of fibres may be due to indirect contact from an indistinguishable source.
Types and proportions of fibres found	The more different types of fibres found and the closer they match the proportions of fibres shed by an item, the more confident one can be that they derive from a common source.
Colour and dye type	Different chemical classes of dye are used in different fibre types. The more distinctive the combination of colour and dyes, the more probative is the match.
Quality of analysis	Some tests are more discriminating than others. A highly discriminating test will increase the significance of matching fibres and the reverse for a poorly discriminating test.
Commonness of fibres	One might anticipate finding a common fibre by chance and unrelated to the particular case. The more unusual the fibre, the more confident one can be that it is due to a true transfer.
Levels of transfer	Fibres being transferred from one item to another is referred to as a 'one-way transfer'. If fibres from the other item are also transferred, this is referred to as a 'two-way transfer'. Two-way transfer considerably increases the significance of the evidence.

Paint

Paint is a complex material with a wide range of potential components but typically includes a coloured pigment suspended in a solvent with other chemical additives. The solvent keeps the paint as a liquid prior to application and the pigment imparts colour. Additives may include chemicals that support the particular method of application, such as spraying, but these evaporate together with the solvent following application. A chemical in the paint known as a binder holds the paint together, forming a hard coating on the surface when dry. Pigments, binders, and other additives derive from a wide range of sources – natural, synthetic, organic, or inorganic – and therefore result in a enormous variation in types and colours of paint. In forensic science, two main classes of paint are encountered: those used for decorative purposes in domestic or commercial premises (architectural paint); and those used for vehicles. Decorative paints are typically encountered in burglaries, and motor vehicle paint is most frequently encountered in 'hit and run' incidents or vehicle crashes. These two types of paint are very different in their composition, and this is reflected in the different methods of analysis and in the interpretation of the evidence involved.

If a burglar uses a tool such as a screwdriver or jemmy to force open a painted window or door, the tool will leave marks on the surface and it is likely that paint will be transferred to the tool. The amount of paint transferred will depend on a range of factors including the force used, the type of surface (metal or wood), and the condition and type of paint involved. Since doors and windows are repainted regularly, there is often more than one layer of paint present. There may also have been more than one attempt to force entry in areas where the paint was a different colour or was in a different condition. It is important that the control sample of the paint from the scene is representative of all of the types of paint present and includes all colours and layers. A

Table 13. Paint examination in a burglary case. This table describes the main stages in the examination of paint

Control sample	Inspect control sample to observe colour and condition of paint without removing from packaging to prevent contamination.
Recovered item	Examine visually and under low-power microscope and individually remove any particles of paint that resemble the control sample.
Comparison	Examine recovered fragments with the control sample using high-power comparison microscopy in white, UV, and polarized light. This takes into account the colours, thickness, and sequence of layers present, and the type of pigment if granular.
Analysis	A wide range of analytical techniques can be used to examine the various components of the paint, including MSP (colour), FTIR (binders, pigments, additives), and X-ray spectroscopy (individual elements such as metals).

summary of the steps involved in the examination of a typical paint case is given in Table 13.

In evaluating the significance of any matching paint, the scientist will take a similar approach to that used with fibres evidence, but the details will be different. In brief, the closer the match and the more uncommon the paint, the stronger will be the value of the evidence. This judgement will take into account the range and types of paint present, the number, colour, and sequence of matching layers, and the significance of any other analyses used in the examination. Finally, the specific circumstances of the case must be considered before a final determination of significance is made.

Glass

Glass is a brittle, hard, transparent amorphous solid. The most common glass around us is *soda glass*, which is a form of calcium silicate. There are other glasses which are manufactured for special purposes e.g. heat resistance, such as *pyrex*, or that contain additional elements, such as boron. Most of the glass encountered in a forensic science laboratory comes from domestic or commercial windows, bottles, and other containers such as drinking glasses. Most domestic glass is made by a process during which it is floated on a bed of molten tin. This gives the glass a very smooth surface which has traces of tin remaining on it. Glass of this type has a fairly standard composition and therefore chemical analysis is of limited value. Another source of glass that is frequently encountered originates from vehicles. In 'hit and run' incidents, fragments of glass from headlamps at the crime scene can be reconstructed and may be able to identify the vehicle type. Glass from the windows of the vehicle will be tempered or toughened, which is why it breaks into the small characteristic fragments found.

If a window is broken by a burglar, some of the glass fragments will be projected backwards and may land on their clothing. Fragments can travel for up to 3 metres and can be recovered from the hair and clothing of individuals who were nearby at the time. The number and size of fragments transferred and retained will depend on the detailed circumstances of the case. Smaller fragments (less than 0.5 millimetres) will transfer more easily and will be retained longer but larger fragments will be lost quickly. Clothing made of fabrics with a smooth surface, such as a cotton shirt, will lose fragments faster than a knitted garment, such as a woollen jumper. Irrespective of how many fragments are transferred, they will be lost very rapidly and are likely to be completely absent within 24 hours.

Examination of glass begins by inspection of the clothing visually and with a low-power microscope. Although the fragments are very small, they can often be seen by the naked eye due to their reflective properties. Each particle found is removed and retained before the item is further searched. To ensure the recovery of all fragments, a process known as 'sweeping' is used. The garment is hung over a large, clean stainless-steel funnel and brushed briskly and systematically with a small stiff-haired brush. This loosens any remaining glass particles, which are then swept down the funnel to collect in a small Petri dish at the bottom. Examination of the fragments by low-power microscopy and a technique known as interferometry can reveal if they have a flat surface and are likely to have come from a window, or a curved surface and are likely to be from a container such as a bottle or drinking glass.

Analysis of the glass invariably commences by measuring its main physical property – refractive index. Refraction takes place when light passes from one medium (such as air) to another (such as glass). The light is slowed down and its path slightly altered. The refractive index is a measure of the ratio of speeds in each medium and varies for different types of glass. This can be measured using a specialist microscope which has a 'hot stage' and is linked to a video system. The glass fragments are placed in a small drop of silicon oil whose refractive index changes according to temperature. The oil is slowly heated by the hot stage until the fragment disappears from view, which is when its refractive index matches that of the oil. Glass can be characterized into groups on the basis of its refractive index, some of which are shown in Table 14. We can see from the table that the result is not definitive and different types of glass overlap in their refractive index. However, together with the physical features of the glass, this can be used to match or eliminate

Table 14. Refractive indices of different glass types encountered in Scotland

Type of glass	Refractive index
Window	1.5080–1.5390
Container	1.5120–1.5230
Headlamp	1.4740–1.4820
Windscreen	1.5130–1.5180

Table 15. Evaluating glass evidence. Factors that are considered in determining the significance of recovered matching glass fragments

Where was the glass found?	Fragments found in hair are more incriminating that those found on shoes, which could be due to contamination from the ground.
Quality of analysis	Physical features and analytical results from refractive index and any further analysis.
Level of transfer	Number and size of fragments recovered.
How common is the glass?	A database of glass fragments can be used to estimate the frequency of the recovered glass in the local environment.
Case details	Specific circumstances including how glass was broken and when clothing was seized, etc.

fragments. Glass also contains a wide range of different chemical elements, such as magnesium, aluminium, potassium, and iron, which can be analysed by a number of specialist techniques. Not all of these minerals are of value in making a comparison and the results can be difficult to interpret.

Given that we have already said that the most commonly encountered glasses are of similar composition, how does one estimate the value of matching fragments? As in other types of trace evidence, a number of factors need to be considered and weighed up, and these are outlined in Table 15.

We have touched on Bayesian evaluation previously, and glass is a good example of how this approach can provide more useful and probative evidence in a case. One way of interpreting glass evidence would be to say that the glass fragments match the source (the broken window) and could have come from it. The court is then left to set this in the context of the case – what other explanations may also account for matching glass being found on someone's clothing. This 'source-level' approach to interpretation does not take into account some very important factors. The first of these is that we know that fragments of glass are not commonly found on clothing and that significant numbers of them are indicative of the person wearing the clothing being close to breaking glass. We also know the circumstances of the case and an experienced scientist can form a view of what she might expect compared to what was actually found. This can then be considered in light of the alleged actions (or activities) which must have taken place if the person committed the crime, that is, broke the window. In such circumstances, we might expect the clothing of that person to have glass fragments on it. We would not expect someone who had no involvement with breaking glass to have significant amounts on them. Taking this approach, we can express the view of how much more likely (or otherwise) it would be that such evidence would be encountered if the accused person had carried out the activity.

Trace evidence is an area of forensic science that can be used in the investigation of many different types of crime. It involves a complex range of analytical procedures and detailed knowledge of the nature of the particular traces involved. We have focused on three of the most common types of trace evidence – fibres, paint, and glass – which exemplify many of the issues involved. Evaluation of trace evidence requires understanding and experience of the particular type of evidence and the detailed circumstances of the case.

Chapter 8
Drugs: identifying illicit substances

A drug is a chemical substance that alters the physiological state of a living organism. Such substances are extensively used in medicine for prevention and treatment of diseases and have a wide range of physiological effects, such as analgesia (pain relief) or anaesthesia. In addition to their medical uses, many drugs are taken solely for pleasure or as a consequence of addiction, and these routinely are the major focus of forensic drugs examinations.

In most countries in the world, production, supply, and possession of certain substances (and chemicals associated with their manufacture) are illegal. In the USA, the Comprehensive Controlled Substances Act (1970) consolidated all prior legislation and classified narcotics and dangerous drugs under five schedules. In Australia, each territory has law dealing with possession, use, distribution, and manufacture of illegal substances. The detailed list of substances differs from country to country, and the law can vary in different parts of the same country, as in Australia. Nevertheless, there are many substances that are commonly restricted around the world, notably cannabis, heroin, and amphetamine. There is also an international list of controlled substances upon which many national controlled lists are based. The common rationale for deeming substances to be illegal is the social harm associated with their misuse. Such judgements are complex and involve social, political, legal (and sometimes religious) dimensions in addition to scientific and medical ones. For example, Schedule I drugs in the USA are considered to have a high potential for misuse, are highly addictive (or cause psychological dependence), and have no currently accepted medical use.

Whatever your views of recreational drug use, the global illicit drugs trade is a major source of human misery. Some grasp of the size of this problem can be gained from estimates of the market size, which is around $20–$25 billion (for cocaine, heroin, cannabis, and synthetic drugs), or roughly equivalent to the global trade in coffee or tea. The total market is estimated at over $100 billion, but only a small number of people make big money. Figure 17 gives some indication of how much money flows through the drugs markets and who benefits from this.

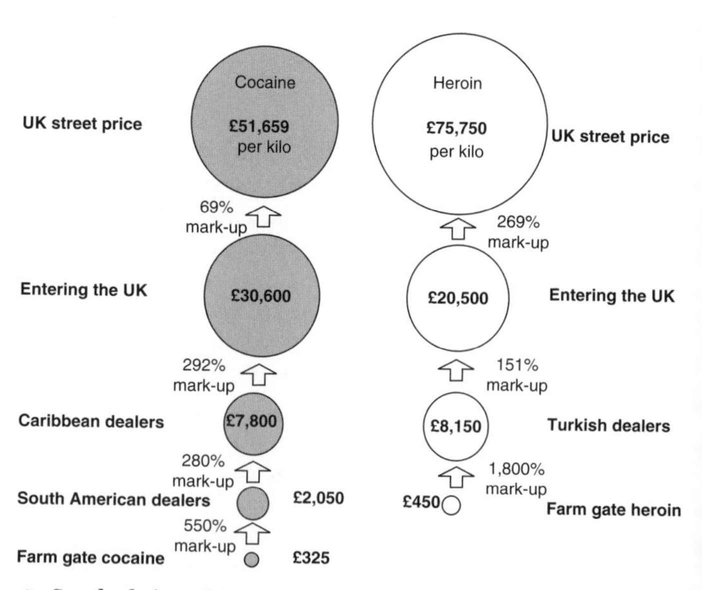

17. Supply chain and financing of drugs supply of cocaine and heroin

Global numbers of users of heroin and cocaine are also increasing, although in most Western countries use of heroin has fallen in the past decade. It follows that forensic examination of drugs in the laboratory is a high-volume activity with one of the largest case loads. Cannabis remains the most widely abused drug globally and accounts for around 70% of the examinations in UK laboratories. However, this volume is likely to decrease slowly for a number of reasons. For a start, technology is now available that can examine drugs at the point of seizure. Although in most countries this is not admissible for court at present, improvements in technology and pressures on criminal justice systems and the police are likely to result in gradual change and convergence in drugs policies around the world. Policies of harm reduction, still controversial in some countries, are finding wider acceptance, and policies towards sellers and traffickers have toughened. Like many areas of criminal justice which are inherently complex, assessment of the effectiveness of policies is difficult due to lack of data and research. Finally, in relation to the impact of drugs on individuals, the issues are also complex. Addiction is not the sole or single source of harm or risk – the lifestyle that follows from drug use can be as big a problem.

Identification of drugs

Identification of illicit substances is central to the investigation of drug crime. But this is not the only aspect that involves forensic examination. Drugs are associated with a wide range of other criminality. Violence, money laundering, prostitution, and use of firearms are often connected with drugs, and much acquisitive crime such as burglary or theft is due to individuals attempting to feed the expensive habit of drug addiction. In many cases of drugs offences there could be no prosecution without scientific evidence. Without identification of the specific substance, the law cannot be enforced. Although in many countries 'fast-track' systems based on field (or presumptive) tests are used to 'identify' drugs, such findings are only usually acceptable in court if the accused admits possessing the specific substance. Where possession is denied or where the charge is supplying or production of drugs, scientific analysis to prove that the seized material is what it is purported to be is normally mandatory. In many instances, the findings of the scientist will determine which crime the accused is charged with and tried for, particularly in distinguishing between possession (that is, for personal use) or supply (trafficking). In relation to the latter, drugs investigations also call on other areas of forensic science, such as fingerprints, DNA, and physical fits, in order to test linkages between individual samples and therefore demonstrate a common source. This may also require drug profiling, analysis of drugs wraps such as 'clingfilm' (food wrap), or contact DNA on packaging.

In this chapter, we consider the main types of drugs available, the relevant law, and the principles of analysis, identification, and quantification. It would be impossible to deal with all of the relevant substances, how they are prepared or synthesized, their analysis and identification, and their legal status even in a book much longer than this, so I will aim to provide the reader with an overview of the range and complexity of forensic drugs examination and some specific examples: cannabis, heroin, and amphetamines.

Analysis of drugs forms a significant part of the overall workload in most forensic science laboratories. Drugs can be analysed by an extraordinarily wide range of methods, and many different methods to analyse the same drug are used in different laboratories. It needs to be emphasized that street drugs are not equivalent to their legal pharmaceutical counterparts. Illicit drugs are prepared or synthesized in sheds, warehouses, fields, and caravans, often in dangerous circumstances using uncontrolled methods and contain many impurities, possibly even toxins. As the drug moves closer to the market, bulking agents are used to cut the drug to increase sales and improve profits. Some of these cutting agents are themselves drugs, for example caffeine or paracetamol, and are deliberately used to fool customers. The purity of street drugs varies enormously and can range from a few per cent to almost pure. The form that a drug takes is also highly variable, as capsules, tablets, powders, resin, oil, herbal material, and so on.

Despite this, in forensic terms analysis and identification in most drugs cases are comparatively straightforward. A combination of visual examination, presumptive testing and standard analytical methodology will rapidly and unequivocally identify most illegal substances. In cases of possession, interpretation of results by the scientist and the courts is straightforward; if an illicit substance is identified, however little is present, the person who was in possession of the material has broken the law. Cases of supply or production in clandestine laboratories can be much more complex. If a person is found in possession of large amounts of a drug, they are likely to be charged with possession and possession with intent to supply. In the strict legal sense (in the UK), simply handing out drugs to your friends at a party could be construed as supply. In circumstances where 'professional' criminals are involved and there are few, if any, witnesses or police surveillance material, the courts will rely on forensic science to provide objective evidence.

Legislation in the UK

There are two main acts in the UK that deal with drugs of abuse: the Misuse of Drugs Act 1971 (MDA), which sets out specific restrictions; and the Misuse of Drugs Regulations 2001 (MDR), which sets out what is allowable, that is, what *can* be done with drugs. MDA introduces the concept of a 'controlled drug' – a substance that, due to its harmful properties, ought not to be freely available to the general public. The act also differentiates between possession of a controlled drug for personal use, possession with intent to supply, and production of controlled drugs, with the legal sanctions becoming increasingly severe. Possession requires only minute amounts and requires no proof of *mens rea* unless the accused can prove that they did not believe or suspect the substance to be a controlled drug.

In addition to specifying which substances are controlled, MDA also classifies substances on the basis of their perceived harm. Class A drugs are considered to be most harmful and include (for example) cocaine, diacetylmorphine, LSD, methadone, methamphetamine, morphine, and psilocin. Class B drugs include amphetamine (unless prepared for injection, in which case it is Class A), codeine, and methaqualone. Examples of Class C drugs include diazepam and ketamine. These are subjective judgements

Table 16. Common illicit drugs: their appearance, origin, and effect (S, synthetic; N, natural; SS, semi-synthetic, i.e. modified)

Drug	Street name	Form	Origin	Effect
Amphetamine	speed, sulphate, wizz	powder, tablet	S	stimulant
3,4-methylenedioxymethamphetamine (MDMA)	ecstasy, E, XTC	tablet	S	hallucinogen
Cannabis (tetrahydrocannabinnol)	marijuana, hash, hashish, grass, resin, oil, weed, skunk, blow, shit	various – resin blocks, dried plant material, or oil	N	mild hallucinogen
Cocaine	snow, charlie, coke	powder	SS	stimulant
Heroin (diacetylmorphine)	H, smack, junk	powder	SS	depressant
Lysergic acid diethylamide (LSD)	acid	tablet	S	hallucinogen

based on a complex assessment of harm carried out by a committee of experts which advises the UK government and which are reviewed from time to time. Nevertheless, these assessments remain controversial. For the most part, there is little connection between drugs policy (and therefore, ultimately, law) and evidence or rationality. The law in the UK is driven almost exclusively by political imperatives. For example, if alcohol were a newly discovered drug, how would it be classified in MDA, given the level of social harm and crime it appears to be associated with? Table 16 provides further information on some of the more common drugs of abuse encountered in the UK, including their street name, typical appearance, and physiological effects.

The Misuse of Drugs Regulations (MDR) sets out what cannot be done with controlled substances, taking into account their value as medicines as opposed to their potential for misuse. MDR controls the manufacture, prescription, and record-keeping of controlled substances.

Cannabis

The cannabis plant *Cannabis sativa* has been used by humans for thousands of years not only as the source of an intoxicating substance, but also of hemp fibres which are used to make ropes and cords. Cannabis is usually encountered in the forensic science laboratory in one of three forms: herbal material (marijuana, grass), resin (hash, hashish), and skunk, a particular form of herbal cannabis which we will consider below. The plant grows in a wide variety of environments but requires high temperatures to provide good yields. Cannabis is the most widely used illegal drug in the UK – in Scotland, 6.3% of 16–59 year-olds reported using the drug in 2004. It is also very commonly used in many other parts of the world. It is often administered by smoking but can also be ingested. The main active ingredient in cannabis is Δ^9 tetrahydrocannabinol (THC), which is classed as a mild hallucinogen. The physiological consequences of chronic use of cannabis are a matter of dispute and, although it is not addictive, it can lead to dependency.

The amount of THC depends on the type of preparation and the quality of the plants. Herbal cannabis has the lowest amount of THC and skunk usually has the highest. Cannabis plants, particularly the flowers, contain a resin and can be compacted into hard blocks (hashish) which have a distinctive appearance and smell. Cannabis resin is readily identifiable in most cases on the basis of a visual examination, which can be confirmed either by the presence of microscopic hairs (trichomes, which are diagnostic) or chemical analysis to establish the presence of THC. Herbal cannabis consists of dried fragments of plant leaves and flowers and occasionally seeds. Since it can resemble dried herbs, it cannot be readily identified visually but it can be by microscopic examination and chemical analysis which detects THC. THC can be easily extracted from the flowers and concentrated in a sticky substance known as 'oil'. The most commonly encountered form of herbal cannabis in the UK is referred to as 'skunk' due to its characteristic odour. Skunk is a form of sinsemilla which consists of the flowering tops of unfertilized plants that have been grown intensively indoors, and it is usually particularly potent.

All forms of cannabis are Class B drugs in the UK, following reclassification in January 2009. Cannabis is a Schedule 1 drug, meaning that it is illegal to produce, possess, and supply without a Home Office licence. Such licences are only granted for research purposes. While possession of seeds is not illegal (they can be found in birdseed), the cultivation of plants is. Allowing premises to be used for the consumption of cannabis is also an offence. Possession of cannabis is still an arrestable offence, though most people over the age of 18 will get a 'cannabis warning' for their first offence.

Heroin

Morphine, diacetylmorphine (heroin), and codeine all derive from the opium poppy (*Papaver somniferum*). Production of opium has been growing since 2006, and 90% of it comes from Afghanistan. Heroin is commonly a white to pale brown powder which is manufactured from the sap of the poppy. It is usually cut with other materials and purity of the active ingredient varies widely. Heroin is usually smoked ('chasing the dragon') or injected, with the latter bringing risk of hepatitis (B and C), infection and HIV. Heroin (and morphine) is highly addictive, the main rationale for its control.

Diacetylmorphine is a synthetic derivative of morphine belonging to the family of drugs known as opiates. Opiates depress brain function, and their main medicinal purpose is sedation and pain relief, but they also produce a feeling of calmness and well-being. Opiates are highly addictive, resulting in dependence and tolerance, which further exacerbate the drug habit and frequently necessitate criminal activity to feed it.

Chronic use of heroin invariably results in poor health, with associated risks of disease, imprisonment, and the destruction of normal social and family life caused by the acquisition and use of the drug. Diacetylmorphine is a Class A, Schedule 2 drug that can only be legally produced, supplied, and possessed under Home Office licence.

Amphetamines

Amphetamine and ecstasy are the most widely abused synthetic drugs in Europe, and European clandestine laboratories are the major suppliers worldwide of ecstasy according to the United Nations Office on Drugs. These drugs belong to a family of hundreds of related compounds which derive from phenethylamine and which can be synthesized by a number of routes. The most commonly encountered of these compounds include methylamphetamine and methylenedioxyamphetamine (MDMA).

Amphetamine as a street drug is usually a powder which ranges in colour from white to yellow or pink. In this form, it can be smoked or snorted, although it can also be ingested. Amphetamine is a stimulant that affects the central nervous system and increases heart rate and blood pressure. It also raises the levels of dopamine and noradrenalin in the blood which cause euphoria. Other effects include a feeling of increased energy, appetite suppression, and reduction in the desire for sleep. Note that it does not provide this extra energy or alter the true need for sleep, which accounts for some of the after-effects of use – tiredness, hunger, and so on. Regular use can lead to addiction. Most amphetamines are Class B, Schedule 2 drugs and therefore can only be possessed, produced, or supplied under Home Office licence.

MDMA is usually encountered in the form of tablets, although powders and capsules are also found, all of which are usually ingested. Many of the tablets have distinctive designs stamped on them. MDMA causes the rapid release of serotonin (5-hydroxytryptamine) and dopamine from nerve endings. The effects of MDMA vary according to dose and the user, but may include increased alertness and sense of energy, sexual arousal, and reduction in desire to sleep. Psychological effects are described as euphoria, increased sense of perception, extraversion, followed by a come-down similar to amphetamine. In acute high doses, MDMA can lead to psychosis.

Ecstasy is a Class A, Schedule 1 drug which has no medical or therapeutic use in the UK. In the UK, any compound which is chemically related to phenethylamine is controlled under the MDR Schedule 1, Part (c).

General aspects of examination

Individual laboratories will use a standard operating procedure for drugs analysis, but the specific technique and methodology will vary from laboratory to laboratory. Whatever technique is used, most examinations will follow a general pattern. The analyst will wear protective clothing to ensure she is not accidentally contaminated by any of the materials examined and to prevent cross-contamination of items. All benches and analytical equipment will be cleaned before the examination and if necessary between examination of different items. Bulk drug samples will be examined in a completely separate area from items which are being examined for traces of drugs. The examination will begin by making detailed records of packaging and sealing of any items to ensure their integrity. In addition to the normal legal procedure for labelling, items for drugs examination are often given a unique reference number which is already pre-printed on the sample bag and is used to refer to the item in any reports.

A visual examination of the item is then carried out. For herbal material and powders, this will include notes of the colour, texture, and weight of the sample as well as its appearance under a low-power microscope. For tablets, their number, colour, shape, and any logos present should be described and recorded. Photographic databases of tablet types are available and can be used to identify and categorize tablets. In non-trace cases, a presumptive test will be carried out on a small sample. Presumptive testing can provide a strong indication that a drug or related substance is present and these are used in the laboratory to screen items. The outcome of the presumptive test will also help determine which analytical methods will be used to further test the sample, since they will indicate what type of substance may be present. Importantly, presumptive tests can in many instances quickly eliminate the presence of controlled substances, saving unnecessary time and analysis.

Table 17. Presumptive tests for drugs. These tests are used to provide an indication of the presence of a controlled substance which can then be confirmed by further analysis

Substance	Marquis Test	Mandelin's Test
Amphetamine	Orange/Orange-brown	Green
Benzoic acid	Purple	
Cocaine	n/a	
Codeine	Blue/Purple	Olive green
Acetylcodeine		Blue
Heroin	Purple	
LSD	n/a	
MDA	Blue/Black-dark brown	
MDEA	Blue/Black-dark brown	
MDMA	Orange/Orange-brown	Green
Diacetylmorphine		Red/Brown
Morphine	Purple	Purple
6 Mono acetylmorphine		Purple
Papaverine		Purple
Noscapine		No reaction
Caffiene		No reaction
Sugar	Pale (lemon) yellow	

Table 17 provides a summary of two of the most widely used presumptive tests with a range of common drugs of abuse. Many of these tests are available in kits and are used by police officers to test materials prior to submission to the laboratory. This allows early elimination of 'innocent' materials as well as the opportunity in some instances for individuals who admit to being in possession of small quantities of certain drugs to be legally processed without the drugs being analysed. Also included in the table are a number of other substances that are not illegal but which may be suspected of being drugs in certain circumstances or may be used as cutting agents.

Analytical methods

As mentioned above, the range of methods that can be used to identify illicit drugs is extremely wide. In this section, we will consider some of the principles involved in the analysis and the generic analytical techniques that may be used. The particular analytical strategy used will depend on the protocols of the individual laboratory and information derived from the initial analysis, including presumptive testing and visual appearance. The primary aim at this stage will be to confirm any presumptive tests and establish the presence of an illicit substance. In some instances, this may be followed by quantification of the amount present which will require either additional analysis or a different methodology from identification. In all cases, whatever analytical method is used, appropriate reference standards of the known substance and related substances will be used and all instruments will be calibrated. At least two, and often three, independent analytical tests are used in order to ensure that a very high standard of identification is maintained that is suitable for the purposes of a criminal prosecution.

When relatively large amounts of sample are present – say a kilogram or hundreds of tablets – not all of the material or all of the tablets will be examined. It is important that an appropriate sampling method is used to ensure the samples removed and analysed are representative of the main item. Failure to do so could provide misleading results. Trace samples present an additional complication since due to the small amount of sample present, presumptive testing will not have been carried out. It is therefore necessary to use a sequence of analytical techniques with minimal use of the sample to carry out an initial chemical classification of the recovered material, prior to full analysis. This may be on the basis of different extraction methods. Some of the methods that can be used to identify drugs have been described elsewhere. The following section describes some of the most common methods which we have not touched on.

Chromatographic methods

Chromatography is an analytical technique which can be used to separate and identify many substances. Generally this involves two phases, a stationary phase and a mobile phase, and the separation process is based on the differential association of components of the mixture of substances, for example a drug sample, with each phase. These differences detected are due to the detailed physical and chemical characteristics of the substances involved. In thin-layer chromatography (TLC), the stationary phase is silica, which is coated on to a plate. Test samples and controls are 'spotted' on to the plate and placed in a tank containing a mixture of solvents. The solvents in this case are the mobile phase, which creep up the plate by capillary action and separate the drugs and other components based on their tendency to remain dissolved or bind to the stationary phase (silica).

In some countries, such as the USA, microcrystalline tests are used to confirm the presence of drugs. In such tests, one of a number of specific reagents is added to the drug which results in the formation of crystals. The shape and colour of the crystals is characteristic of the type of drug. Gas chromatography uses the same principles as thin-layer chromatography to separate gases. To use this method, the sample is vaporized and passed through a very fine long tube (the column) in a carrier gas such as hydrogen. Inside the tube is a solid (e.g. keiselguhr) or a non-volatile liquid which acts as the stationary phase. The components in the vaporized sample are separated due to their differing tendencies to bind to the stationary phase or remain in the mobile phase. Generally speaking, components are identified by how long they take to pass through the column, which is referred to as their 'retention time'. The nature of the stationary phase, mobile phase, column length, temperature, and so on all have a bearing on how effectively gas chromatography separates components, and these can be optimized in advance for each drug or class of drugs. When the substance leaves the column, it can be detected by a variety of means, with the most common method being flame ionization. Gas chromatography is more precise and reproducible than thin-layer chromatography but also has its limitations: not all substances can be easily converted into gases and some substances cannot be separately resolved.

High-performance liquid chromatography can also be used to identify drugs on the basis of their retention time. It has the advantage that samples do not require pre-treatment (such as vaporization), although the substance must be soluble in a number of solvents. This technique also has the benefit that analysis can be automated and it can be used to quantify samples.

Trafficking and supply of illicit drugs

Much of this chapter has been about the identification of substances. In cases of trafficking, further work in addition to identification is required. In many instances, this additional evidence is supplied by the police from investigation or surveillance, especially in large-scale importation when hundreds of kilos may be involved. In some cases, no such evidence is available, or what evidence is available is insufficient to support a prosecution. Where many drugs samples are recovered from different sources and different parts of the supply chain, detailed analysis of the drugs and associated packaging materials can be used to establish linkages between samples.

There are a variety of means by which this can be done. Where individual 'deals' are wrapped in paper, physical fits may be found that link batches. Where magazines or newspapers are used, these can be used to link samples on the basis of similar characteristics or analysis of the paper and inks. Drugs wrapped in paper or in plastic bags can also be examined for fingerprints from suspected persons. Paper, plastic bags, and food wrap ('clingfilm') used to wrap drugs can be examined for DNA traces. Striations on plastic bags which are a result of manufacturing

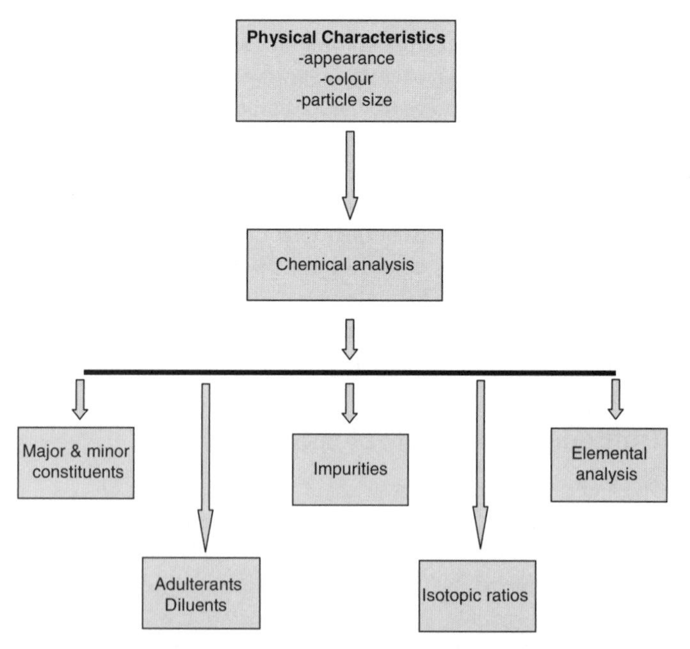

18. Profiling of drugs. This can be carried out by a number of means but requires extensive and detailed work

processes can be used to link batches of plastic bags. Detailed analysis of the drugs and other chemicals present can be used to profile samples to determine potential linkages as outlined in Figure 18.

Profiling of drug samples requires an extensive and detailed examination of the materials seized. Although physical characteristics of the material can be relatively straightforward to compare, chemical analysis of the constituents is complex, time-consuming, and can be difficult to interpret. Such examinations are more frequently used for intelligence purposes to link potential sources of supply than as evidence in court.

Analysis of illicit substances is a routine aspect of the forensic laboratory, involving a wide range of analytical techniques depending on the particular substances involved. The detailed analysis is usually preceded by a simple presumptive test which gives an indication of the drug involved. Although the interpretation of findings is dependent on the law and linked to the particular country where the analysis is carried out, many countries restrict the same substances due to the harm they cause to individuals and the association of the drugs trade with other forms of criminality.

Chapter 9
Science and justice

> The enormous conceptual change that [scientific] thinking require[s]d shows that science is not just about accounting for the 'unfamiliar' in terms of the familiar. Quite the contrary: *science often explains the familiar in terms of the unfamiliar.*
>
> Lewis Wolpert, *The Unnatural Nature of Science*

> More and more the problem of expert qualification and the risk of biased scientific evidence appear to stem from the institutional demands and limitations of criminal proceedings rather than reflecting the inadequacies of scientific method or failure of individual experts.
>
> Paul Roberts and Christine Willmore, *The Role of Forensic Science Evidence in Criminal Proceedings*

The defining feature of forensic science is its relationship with the law. In this chapter, we come to what is typically the final stage of a criminal inquiry – the trial. This involves an encounter between science and law, and we will reflect on the issues this may raise. In doing so, we will explore the nature of science and law and the implications of their different worldviews for how forensic science is used, develops, and is constrained. Science is based on observing the external world. It has no particular view of how the world ought to be, but seeks to describe it on the basis of empirical observations and the development of predictive models. These models are continually retested and modified on the basis of experiments which use universal methodologies. Central to these methodologies is the use of statistical probability to describe the level of uncertainty in any set of observations. Science is the same wherever it is done and is unquestionably the best means we have of understanding the physical world. Scientific models do not always concur with our commonsense view of the world; they can be unfamiliar, even counterintuitive. In fact, Lewis Wolpert, the embryologist and well-known science writer quoted above, considers this to be one of the defining features of science: it not only explains the unfamiliar in terms of the familiar, but sometimes explains the familiar in new ways that we could not have foreseen.

Law proceeds by its own authority, by the power of statute or case law, with limited reference to any external authority. Different legal systems operate in different ways, and there is no universal law comparable to universality of science. In fact, the law is local, at the level of the nation state or provincial administration. The legal world can be divided into two main families of criminal law: inquisitorial and adversarial. In adversarial legal systems (the USA, the UK and its former colonies and dominions, Australia, Canada, New Zealand), legal outcomes are determined on the basis of a highly structured argument. Only certain types of facts and information can be used in such arguments as these need to be allowable (admissible) under the rules of evidence. The rules of evidence are an essentially *ad hoc* collection of mainly exclusionary directives derived from historical practice. Inquisitorial systems are less restrictive about the evidence they allow and more interested in the potential value of the evidence in determining the outcome to the case. In general, inquisitorial systems get nearer to the 'truth' than adversarial systems. Unlike science, which uses a consistent methodology and standards, the law uses different procedures depending on the circumstances. In criminal cases, the prosecution has to prove 'beyond reasonable doubt' that the defendant is guilty, but in civil cases the burden of proof is on the balance of probabilities. Such determinations are frequently made by juries, and the rationale for such judgements is 'common sense'. It is difficult to imagine two systems of knowledge that go about their businesses so differently, and it should come as little surprise that there are inevitable conflicts between science and law about who is right and whose knowledge has most authority.

These issues most frequently arise when expert evidence is involved. The law has recognized for hundreds of years the need to use special types of witnesses called 'expert witnesses' to cover areas of evidence about which the courts lack knowledge and expertise in the matters before them. Expert witnesses differ from ordinary witnesses in a number of ways, but primarily in their right to give opinions in evidence. Ordinary witnesses must restrict their evidence to facts – what they observed or heard – but are not allowed to interpret these facts, as this is a matter for the jury. Expert witnesses can express an opinion about the meaning and significance of facts – whether an item is a firearm, how a fire started, how a weapon came to be bloodstained. However, they are only entitled to express opinions within their particular area of expertise.

This is a particularly difficult issue, which can be illustrated by a recent miscarriage of justice in the UK. In 1999, the solicitor Sally Clark was convicted of the murder of her two infant children. The expert witness, Sir Roy Meadow, then Professor of Paediatrics and Child Health at Leeds University, gave evidence at her trial which was highly influential in her conviction. He testified on a range of medical matters that were presumably within his area of expertise, but he was also allowed by the court to give expert opinion on the probability of two child deaths in the same family. He stated that this probability was 1 in 73 million. There were no statistical experts involved in the trial and no significant cross-examination on this point. Meadow derived this figure from the frequency of a single child death in a family (1 in 8,500), which he multiplied to account for both deaths to obtain 1 in 73 million. His mistake should now be clearly evident. In Chapter 5, we encountered how the significance of DNA profiles are estimated on the basis of the combination of genotype frequencies involved. Such frequencies can only be multiplied if they derive from independent events, but the deaths of two genetically related individuals (the children in this case) cannot be treated as independent events. The subsequent appeal which released Clark also resulted in the review of around 5,000 other cases.

In the UK, the role and responsibilities of the expert witness are set out in case law, although there is also some legislation and practice guidance. Case law requires an expert to have 'expertise', but does not set out criteria by which this is to be judged or established. There is no need for experts to be formally qualified, indeed there is case law that warns against being over-impressed by the qualifications of expert witnesses, something that perhaps applied in the Clark case. The law also carefully prescribes the ambit of expert witnesses and how they ought to interact with the courts. The main case on this issue is worth quoting:

> [The role of the expert is] to furnish the judge with the necessary scientific criteria for testing the accuracy of their conclusions, so as to enable the judge or jury to form their own independent judgement by the application of these criteria to the facts proved in evidence.

This carefully crafted judgement was set out by Lord Justice Cooper in a Scottish case (*Davie* v. *Edinburgh Magistrates*) in 1953. In effect, this amounts to an expectation that the expert witness will educate the court in each case on the specialist subject in hand – analysis of drugs, DNA profiling, paint comparison. This is a tall order in a classroom with willing volunteers such as students; it is almost impossible to meet this standard in a courtroom where the questioning is under the control of two lawyers who have opposing views about the significance of the evidence, and seek to convince the jury of their interpretation. The following example illustrates what can be achieved by tactical cross-examination:

Defence counsel: 'Can you age hair?'

Witness: 'No'

Defence counsel: 'Are you sure?'

Witness: 'I am not aware of any means by which you can age biological materials.'

Defence counsel: 'Have you done any research which proves that you cannot age hair?'

Witness: '. . . No'

Defence counsel: 'Have you read any published papers which state that you cannot age hair?'

Witness: 'I am not aware . . .'

Defence counsel: 'You are not aware of any evidence which states that you cannot age hair . . .'

The example dates from the 1980s, but the principle it illustrates is still valid today. By careful questioning and control of the witness, defence counsel elicits a 'fact' from the witness (the possibility that hair can be aged cannot be excluded) which is untrue and which no one in the expert community believes to be true. Furthermore, the implication is left that the expert should perhaps have done her homework rather better and ought to have tried to age the hairs in the case. Most experts dislike this type of sophistry, but it is quite common in the courtroom. An experienced expert can respond in turn with carefully worded and qualified answers in what can become a war of words, but it rarely gets to the truth. Since it is not possible to carry out research on these issues, we cannot know the impact of such exchanges on juries. Such issues are not confined to cross-examination. This is not an environment that supports the level and quality and communication required of such important issues, and the quote from Roberts above appears to bear this out. Even the physical layout of the courtroom can intrude. The witness generally faces the questioning advocates, as is appropriate, but typically has to turn through 90 degrees to make eye contact with the jury who will evaluate her evidence, and the judge is frequently behind the witness. Those experts I know who have given evidence in inquisitorial systems consider them to be far more conducive to effective communication of evidence.

In the USA, expert evidence is subject to two tests before it can be deemed admissible. The first of these (*Frye* v. *The United States*, 1923) seeks to establish the reliability of the potential evidence by asking if it has 'gained general acceptance in the particular field in which it belongs'. This is sound in principle but difficult to apply in practical terms. How does one establish 'general acceptance', and furthermore, what about new technology which by definition cannot have attained general acceptance? A second test is used in the USA which particularly addresses the issue of science. In *Daubert* v. *Merrell Dow Pharmaceuticals Inc.* (1993), it was determined that certain expert testimony must be based on 'scientific knowledge'. The determination of what is and what is not science is more difficult than it appears since science has no single agreed definition. These considerations notwithstanding, both *Frye* and *Daubert* attempted to deal with the issue of expert evidence by setting standards for admissibility and reliability. The approach brings some clarity and shared understanding of this complex situation to lawyers and scientists, of the standards expected of expert witnesses and the restrictions that may be applied to evidence which does not meet the standard. There are no such declared standards in the UK at present, although the matter is under review by the Law Commission in England and Wales at the time of writing. The courts recognize the need for expert witnesses who have knowledge not in possession of the law. Yet there is a fundamental contradiction, since it is individual courts on an incremental case-by-case basis that determine who is to be an expert witness and what constitutes expert evidence.

The relationship between science and law is a complex one. Most lawyers are ignorant of science and most scientists ignorant of law.

The courtroom is a complex environment which does not readily support the level or quality of communication that such evidence merits. Given that the amount of scientific evidence entering legal systems is higher than ever before, the development of an effective relationship between science and law is essential to ensure science continues to contribute to criminal justice.

Afterword

> . . . the umbrella term 'forensic science' embraces a set of intensely practical disciplines to which the paradigm of pure scientific enquiry cannot readily be applied.
>
> Paul Roberts and Christine Willmore, *The Role of Forensic Science Evidence in Criminal Proceedings*

I have tried to provide the reader with some understanding of forensic science, its value, limitations, and potential. As Roberts and Willmore suggest above, forensic science does not conform readily to our general expectations or requirements of science. It is messy, conceptually and practically; it deals with body fluids and body parts, explosions, burned-out buildings, and shattered fragments from a myriad of sources, which it attempts to piece together in some meaningful way within the constraints of the legal framework in which it is applied. The boundaries of forensic science are unclear, or at least contested, its evidence base is weak in some areas and open to challenge. It is in turn lauded and condemned by the press, politicians, and the public as befits their mood, and it can be prone to exploitation by bogus experts.

Forensic science makes a unique contribution to criminal justice, despite these infirmities, providing answers not achievable by any other means and to a standard unmatched by any other type of evidence. Its contribution remains attenuated because it is indirectly applied by practitioners (police and lawyers) who have limited understanding of its potential, and by scientists who fail to grasp the legal or investigative significance of information in their possession. Recognition of this interdependence of science, law, and policing, of the importance of shared knowledge and improved communication, is the key to a more effective contribution by forensic science to criminal justice.

Further reading

A. Bell, J. Swenson-Wright, and K. Tybjerg (eds.), *Evidence* (Cambridge: Cambridge University Press, 2008). An edited volume on the nature of evidence in a wide range of contexts but with a chapter on statistics and the law which draws on the Sally Clark/ Meadows case.

S. Bell, *Forensic Chemistry* (New Jersey: Pearson Prentice-Hall, 2006). A good introduction to a number of important areas of forensic chemistry. Requires some understanding of chemistry.

J. M. Butler, *Forensic DNA Typing: Biology, Technology, and Genetics of STR Markers* (London: Academic Press, 2005). The standard DNA reference which requires significant technical knowledge of the subject.

J. Fraser and R. Williams (eds.), *The Handbook of Forensic Science* (Cullompton: Willan, 2009). An edited volume which covers all of the areas in this book as well as social, legal, economic, and political aspects of forensic science. Most contributors are recognized international experts in their subjects.

W. Goodwin, A. Linacre, and S. Hadi, *An Introduction to Forensic Genetics* (Chichester: Wiley, 2007). A basic and accessible introduction to DNA profiling and population genetics.

M. H. Houck and J. A. Siegel, *Fundamentals of Forensic Science* (Boston: Academic Press, 2006). An introductory text from an American perspective for those with some knowledge of science.

National Research Council, *Strengthening Forensic Science in the United States: A Path Forward* (Washington, DC: National Academies Press, 2009). A detailed review of forensic science in the

USA which raises many important issues for the USA in particular as well as other countries.

R. Williams and P. Johnson, *Genetic Policing: The Use of DNA in Criminal Investigations* (Cullompton: Willan, 2008). A comprehensive account of the development and use of DNA databases throughout the world from a sociological perspective.

Index

62. Schopenhauer
63. The Russian Revolution
64. Hobbes
65. World Music
66. Mathematics
67. Philosophy of Science
68. Cryptography
69. Quantum Theory
70. Spinoza
71. Choice Theory
72. Architecture
73. Poststructuralism
74. Postmodernism
75. Democracy
76. Empire
77. Fascism
78. Terrorism
79. Plato
80. Ethics
81. Emotion
82. Northern Ireland
83. Art Theory
84. Locke
85. Modern Ireland
86. Globalization
87. The Cold War
88. The History of Astronomy
89. Schizophrenia
90. The Earth
91. Engels
92. British Politics
93. Linguistics
94. The Celts
95. Ideology
96. Prehistory
97. Political Philosophy
98. Postcolonialism
99. Atheism
100. Evolution
101. Molecules
102. Art History
103. Presocratic Philosophy
104. The Elements
105. Dada and Surrealism
106. Egyptian Myth
107. Christian Art
108. Capitalism
109. Particle Physics
110. Free Will
111. Myth
112. Ancient Egypt
113. Hieroglyphs
114. Medical Ethics
115. Kafka
116. Anarchism
117. Ancient Warfare
118. Global Warming
119. Christianity
120. Modern Art
121. Consciousness
122. Foucault
123. The Spanish Civil War
124. The Marquis de Sade
125. Habermas
126. Socialism
127. Dreaming
128. Dinosaurs
129. Renaissance Art
130. Buddhist Ethics
131. Tragedy
132. Sikhism
133. The History of Time

134. Nationalism
135. The World Trade Organization
136. Design
137. The Vikings
138. Fossils
139. Journalism
140. The Crusades
141. Feminism
142. Human Evolution
143. The Dead Sea Scrolls
144. The Brain
145. Global Catastrophes
146. Contemporary Art
147. Philosophy of Law
148. The Renaissance
149. Anglicanism
150. The Roman Empire
151. Photography
152. Psychiatry
153. Existentialism
154. The First World War
155. Fundamentalism
156. Economics
157. International Migration
158. Newton
159. Chaos
160. African History
161. Racism
162. Kabbalah
163. Human Rights
164. International Relations
165. The American Presidency
166. The Great Depression and The New Deal
167. Classical Mythology
168. The New Testament as Literature
169. American Political Parties and Elections
170. Bestsellers
171. Geopolitics
172. Antisemitism
173. Game Theory
174. HIV/AIDS
175. Documentary Film
176. Modern China
177. The Quakers
178. German Literature
179. Nuclear Weapons
180. Law
181. The Old Testament
182. Galaxies
183. Mormonism
184. Religion in America
185. Geography
186. The Meaning of Life
187. Sexuality
188. Nelson Mandela
189. Science and Religion
190. Relativity
191. History of Medicine
192. Citizenship
193. The History of Life
194. Memory
195. Autism
196. Statistics
197. Scotland
198. Catholicism
199. The United Nations
200. Free Speech
201. The Apocryphal Gospels

202. Modern Japan
203. Lincoln
204. Superconductivity
205. Nothing
206. Biography
207. The Soviet Union
208. Writing and Script
209. Communism
210. Fashion
211. Forensic Science

LOGIC

A Very Short Introduction

Graham Priest

Logic is often perceived as an esoteric subject, having little to do with the rest of philosophy, and even less to do with real life. In this lively and accessible introduction, Graham Priest shows how wrong this conception is. He explores the philosophical roots of the subject, explaining how modern formal logic deals with issues ranging from the existence of God and the reality of time to paradoxes of self-reference, change, and probability. Along the way, the book explains the basic ideas of formal logic in simple, non-technical terms, as well as the philosophical pressures to which these have responded. This is a book for anyone who has ever been puzzled by a piece of reasoning.

> 'a delightful and engaging introduction to the basic concepts of logic. Whilst not shirking the problems, Priest always manages to keep his discussion accessible and instructive.'
>
> **Adrian Moore, St Hugh's College, Oxford**

> 'an excellent way to whet the appetite for logic. . . . Even if you read no other book on modern logic but this one, you will come away with a deeper and broader grasp of the *raison d'être* for logic.'
>
> **Chris Mortensen, University of Adelaide**

www.oup.co.uk/isbn/0-19-289320-3

MEDICAL ETHICS

A Very Short Introduction

Tony Hope

Issues in medical ethics are rarely out of the media and it is an area that has particular interest for the general public as well as the medical practitioner. This short and accessible introduction provides an invaluable tool with which to think about the ethical values that lie at the heart of medicine. Tony Hope deals with the thorny moral questions such as euthanasia and the morality of killing, and also explores political questions such as: how should health care resources be distributed fairly?

Each chapter in this book considers different issues including; genetics, modern reproductive technologies, resource allocation, mental health, and medical research.

'...engrossing taster' – **Paul Nettleton, Guardian Life**

http://www.oup.co.uk/isbn/0–19–280282–8

MOLECULES

A Very Short Introduction

Philip Ball

The processes in a single living cell are akin to that of a city teeming with molecular inhabitants that move, communicate, cooperate, and compete. In this Very Short Introduction, Philip Ball uses a non-traditional approach to chemistry, focusing on what chemistry might become during this century, rather than a survey of its past

He explores the role of the molecule in and around us - how, for example, a single fertilized egg can grow into a multi-celled Mozart, what makes spider's silk insoluble in the morning dew, and how this molecular dynamism is being captured in the laboratory, promising to reinvent chemistry as the central creative science of the century.

> 'Almost no aspect of the exciting advances in molecular research studies at the beginning of the 21st Century has been left untouched and in so doing, Ball has presented an imaginative, personal overview, which is as instructive as it is enjoyable to read.'
>
> **Harry Kroto, Chemistry Nobel Laureate 1996**

> 'A lucid account of the way that chemists see the molecular world . . . the text is enriched with many historical and literature references, and is accessible to the reader untrained in chemistry'
>
> **THES, 04/01/2002**

http://www.oup.co.uk/isbn/0–19–285430–5